CHIMNEY AND STACK INSPECTION GUIDELINES

EDITED BY
Bernhardt H. Hertlein

Published by the American Society of Civil Engineers

Library of Congress Cataloging-in-Publication Data

Chimney and stack inspection guidelines / edited by Bernhardt H. Hertlein.
 p. cm.
 Includes bibliographical references and index.
 ISBN 0-7844-0693-6
 1. Chimneys--Inspection. I. Hertlein, Bernhardt H.

TH4591.C45 2003
697'.8'--dc22   2003050040

American Society of Civil Engineers
1801 Alexander Bell Drive
Reston, Virginia, 20191-4400

www.pubs.asce.org

Any statements expressed in these materials are those of the individual authors and do not necessarily represent the views of ASCE, which takes no responsibility for any statement made herein. No reference made in this publication to any specific method, product, process, or service constitutes or implies an endorsement, recommendation, or warranty thereof by ASCE. The materials are for general information only and do not represent a standard of ASCE, nor are they intended as a reference in purchase specifications, contracts, regulations, statutes, or any other legal document. ASCE makes no representation or warranty of any kind, whether express or implied, concerning the accuracy, completeness, suitability, or utility of any information, apparatus, product, or process discussed in this publication, and assumes no liability therefore. This information should not be used without first securing competent advice with respect to its suitability for any general or specific application. Anyone utilizing this information assumes all liability arising from such use, including but not limited to infringement of any patent or patents.

ASCE and American Society of Civil Engineers—Registered in U.S. Patent and Trademark Office.

Photocopies: Authorization to photocopy material for internal or personal use under circumstances not falling within the fair use provisions of the Copyright Act is granted by ASCE to libraries and other users registered with the Copyright Clearance Center (CCC) Transactional Reporting Service, provided that the base fee of $18.00 per article is paid directly to CCC, 222 Rosewood Drive, Danvers, MA 01923. The identification for ASCE Books is 0-7844-0693-6/03/ $18.00. Requests for special permission or bulk copying should be addressed to Permissions & Copyright Dept., ASCE.

Copyright © 2003 by the American Society of Civil Engineers.
All Rights Reserved.
Library of Congress Catalog Card No: 2003050040
ISBN 0-7844-0693-6
Manufactured in the United States of America.

Acknowledgments

Preparation of a set of guidelines such as these is an onerous task, and one that involves many people. It is a commitment not to be undertaken lightly, yet it was gladly undertaken by a volunteer group of the industry's most knowledgeable professionals, who gave generously of their time and efforts. Not only did these individuals give much of their personal time to the project, but they somehow persuaded their employers to donate time and money for attendance at committee meetings, and innumerable fax, FedEx and email transactions! I think it is only fair, then, to acknowledge the great effort contributed by the following individuals and the companies that employ them. The members of the ASCE task committee for the preparation of these guidelines were:

David Bednash*
Bednash Consulting, Inc, Roselle, IL

Arun K. Bhowmik*
Hamon Custodis, Somerville, NJ

David Bird*
Pullman Power Products, Kansas City, MO

Victor Bochicchio*(Co-chairman)
Hamon Custodis (Formerly Zurn Balcke-Durr) Somerville, NJ

William Brannen
Florida Power & Light Co, Juno Beach, FL

Richard Burkee
Carolina Power & Light, Raleigh, NC

John J. Carty
R&P Industrial Chimney Co, Nicholasville, KY

Joseph DeMartino
Hamon Custodis, Somerville, NJ

William Foster
Carolina Power & Light, Raleigh, NC

Rick Harris*
Tennessee Valley Authority, Chattanooga, TN

Bernard Hertlein*(Co-chairman)
STS Consultants, Ltd, Vernon Hills, IL

Thomas Joseph
Florida Power & Light Co, Juno Beach, FL

Beth Kasperski
Florida Power & Light Co, Juno Beach, FL

Erick N. Larsen
Pacific Gas & Electric Co, San Francisco, CA

Kenneth Mixer
Sargent & Lundy, Chicago, IL

Robert Porthouse*
Chimney Consultants, West Lebanon, NH

Rodney K. Simonetti
Parsons Power Group, Reading, PA

James L. Thorp
PSI Energy, Plainfield, IN

Timothy Topor*
Gibraltar Chimney Int'l, LLC, Tonawanda, NY

Raymond M. Warren Jr.*
Warren Environmental, Inc, Atlanta, GA

(* denotes contributing author)

Thanks are also due to Ron Schneider and John Hill, who, as former and current chairmen, respectively, of the ASCE Fossil Power Committee, supported the task committee in its work. As the reader can see, this group represents the experience of chimney designers, constructors, owners, inspectors, and repair specialists. Thank you, lady and gentlemen—I believe that the industry will truly benefit from our efforts.

Bernard H. Hertlein
Editor

Contents

Scope 1

1.0 Introduction 3
  1.1 Overview 3
  1.2 Classes of Inspection 4
  1.3 Inspection Intervals 8
  1.4 Common Forms of Degradation 11

2.0 General Inspection Methods and Procedures 14
  2.1 Rationale Behind Inspection Procedures 14
  2.2 Site Planning an Inspection 14
  2.3 Visual Inspection, Photography, and Documentation 17
  2.4 Measurement and Instrumentation 18
  2.5 Nondestructive Testing 18
  2.6 Sampling and Laboratory Testing of Concrete and Masonry 27
  2.7 Sampling and Laboratory Testing of Metals and FRP 28
  2.8 Dynamic Monitoring and Modal Analyses 29
  2.9 Typical Inspection Program Specifications 31
  2.10 Other Information to Include in Bid Request Documents 31
  2.11 Record Keeping/Report Archives 31
  2.12 Minimum Inspection Team Qualifications and Safety Training 32

3.0 Concrete Shells 34
  3.1 General Information 34
  3.2 Construction Materials 35
  3.3 Degradation/Modes of Failure 35
  3.4 Inspection Programs 37

4.0 Brick Shells 45
  4.1 General Information 45
  4.2 Construction Materials 45

4.3   Degradation/Modes of Failure 47
4.4   Inspection Programs 47

5.0   Steel and Alloy Stacks 55
5.1   General Information 55
5.2   General Descriptions 55
5.3   Construction Materials 56
5.4   Degradation/Modes of Failure 57
5.5   Inspection Programs 58

6.0   Brick Liners 66
6.1   General Information 66
6.2   Construction Materials 67
6.3   Degradation/Modes of Failure 67
6.4   Inspection Programs 68

7.0   Steel and Alloy Liners 71
7.1   General Description 71
7.2   Construction Materials 73
7.3   Degradation/Modes of Failure 73
7.4   Inspection Programs 75

8.0   Fiberglass Reinforced Plastic (FRP) Liners and Stacks 78
8.1   General Information 78
8.2   Construction Materials 81
8.3   Degradation/Modes of Failure 81
8.4   Inspection Programs 83

9.0   Coatings and Linings 89
9.1   Coating Manuals and Standards 89

10.0   Appurtenances 91
10.1   Inspection of Appurtenances 91

11.0   Glossary of Terms 93

Appendix A—Sample Inspection Report Specification 96

Appendix B—Sample Checklists and Forms 98

Appendix C—Example of Developed Plan of a Concrete Utility Chimney 117

Bibliography and Standards References 118

Index 121

CHIMNEY AND STACK INSPECTION GUIDELINES

SCOPE OF THIS DOCUMENT

This document addresses the inspection of chimneys and stacks. These guidelines are intended to aid owners in determining what level of inspection is appropriate to a particular chimney and to provide common criteria so that all parties involved have a clear understanding of the scope of the inspection and the end product required.

Each chimney or stack is a unique structure, subject to both aggressive operating and natural environments, and degradation over time. Such degradation may be managed via a prudent inspection program followed by maintenance work on any equipment or structure determined to be in need of attention. The purpose of this document is to provide a useful guidance tool and set of references.

Visual inspection procedures are inherently subjective. Photographs of the same part of a chimney or stack can appear different if taken under different light conditions. Quantification of observations is therefore important. The use of measurements, nondestructive tests and laboratory assessment of material samples can greatly enhance the objectivity of the inspection.

Proper inspections can establish a maintenance database provided that the inspection methods ensure that accurate and repeatable data are obtained. Design drawings and previous inspection reports, when available, should be provided to the personnel selecting the inspection program, and those performing the inspection. Data taken must include ambient site conditions, test procedures and test locations, with all these factors well documented.

This document addresses the following:

- Three distinct classes of inspection, and their use
- Inspection techniques that can minimize subjectivity
- Inspection programs and reporting formats
- A glossary of terms common to chimney and stack inspection and maintenance

This guide recognizes that the individuals often responsible for chimney maintenance at a utility or industrial facility may be given that duty in addition to other responsibilities out of commercial necessity, and may, in fact, have no prior experience in chimney construction, operation, or maintenance. The glossary is included for readers that are relatively new to the business of chimney inspection and maintenance and who may therefore be unfamiliar with some of the terms used in this manual.

Where trade names of equipment or test methods are mentioned in this guide, they are clearly identified by italic letters. Use of trade names does not imply any endorsement or preference for the named item. The name is used only to provide a well-known example of a generic type of equipment or test method, and avoid possible confusion with similarly named but different products or services.

1.0 INTRODUCTION

1.1 Overview

Generally in the various industries concerned with the construction, operation, and maintenance of structures needed to vent the by-products of combustion, the term "chimney" refers to a reinforced concrete or brick shell construction, and the term "stack" refers to steel or fiber-reinforced plastic (FRP) shell construction. However, since "stack" is a common contraction for both the terms "chimney stack" and "smokestack", the term "chimney" is used synonymously for both herein. Additional definitions are shown in Figures 1, 2, and 3, in the following pages.

Utility and industrial chimneys are usually exposed to severe environmental conditions (e.g. chemical attack, high operating temperatures, extremes of heat and cold, rapid heating, vibration and earthquake, high wind loads). They are also often subjected to operating conditions that change over time (e.g. fuel switching, duty cycling, structural modifications, and changes to environmental control processes). Often, owners of facilities may not be familiar with the effects of environmental and operating conditions, or changes in them, on the behavior of these structures, and do not recognize early signs of distress. Repair activities involving significant and unplanned expenditure are often triggered by conditions that may have been mitigated if inspection and maintenance had been performed in a timely manner.

For the purposes of this guide, the terms "Inspection" and "Condition assessment" are not synonymous. "Inspection" refers to observations and measurements made to document the existing conditions on a chimney. "Condition assessment" refers to determining the significance of the recorded conditions in terms of chimney stability, remaining service life, and possible maintenance or repair needs. This guide does not attempt to make recommendations for repair or renovation, which require in-depth structural review and evaluation. Such recommendations should be based on the observations and experience of the trained plant personnel or specialist firm performing the inspection. This document focuses on the inspection technology and collection of data necessary to make informed decisions.

A comprehensive inspection program can help ensure long-term structural integrity, minimize maintenance expenditure, and eliminate unplanned outages, resulting in significant cost savings. In the case of retired chimneys, structural stability can be affected by weathering, flue gas from neighboring chimneys and other environmental factors. For personnel and plant safety, regular inspection of retired chimneys is therefore as important as it is for active chimneys.

Even when the need for periodic inspection is recognized, many facility owners do not have a good basis for determining what is required. Inspection programs can range from simple safety inspections of ladders, platforms, and stairs, to thorough structural evaluations using state-of-the-art dynamic monitoring and computerized surface mapping. In addition, the design of chimneys is a highly specialized field. Most engineering and maintenance personnel are not familiar with unique design considerations associated with chimneys. As a result, it can be very difficult for a facility owner to determine what constitutes a well-designed inspection program. The purpose of this guide is to provide some assistance by defining a set of good engineering practices that owners can use as a basis for implementing an adequate inspection and maintenance program.

This guide is not intended to be an exclusive specification. Other programs developed by knowledgeable and qualified individuals can be just as effective. The intent of this guide is to provide a basis to be used when specific or customized programs are not available or warranted.

This guide is not intended as a primer for inspectors with no previous chimney experience. For the purposes of this guide, it is assumed that the personnel performing the inspection procedures recommended herein are experienced in chimney inspection, and in the performance of the specialized measurements and tests recommended, where applicable (See section 2.12).

1.2 Classes of Inspection

The scope and frequency of an inspection program will vary with a number of factors, such as age, outage schedule, initial operating conditions, changes in operating conditions, visible degradation, and the importance and safety of the chimney. The intervals recommended in this manual are for guidance only, and may need to be substantially reduced, depending on site conditions.

It is important to bear in mind not only that experienced personnel should perform the inspection, but also that the appropriate class of inspection and inspection program should be determined by a qualified person, preferably a professional engineer experienced in the design and construction of chimneys.

As a guide for facility owners and engineers, the committee has defined the scope of recommended classes of inspection. More detailed information about various chimney arrangements, types of linings, common forms of degradation, and the inspection or monitoring methods classified in the following paragraphs can be found in subsequent chapters.

It must be understood that the following classes of inspection are implemented on a periodic basis, and should be supplemented by good in-plant housekeeping procedures. Such procedures should include regular (at least monthly) walk-around

CHIMNEY AND STACK INSPECTION GUIDELINES 5

visual surveys of the chimney and working platforms to observe signs of degradation such as fallen debris, new cracks, spalled or crumbling concrete or brickwork, corrosion or wear of exposed metal and appurtenances, or blistering or discoloration of Fiberglass Reinforced Plastic (FRP) sections.

These walk-around inspections also provide the opportunity to prevent the accumulation of maintenance debris and waste on platforms, catwalks, and adjacent roofs. Such accumulations of debris present a trip-and-fall hazard for workers, and are a potential falling hazard in strong winds or heavy rain.

1.2.1 CAUTION – Health Hazards:

Many chimneys may contain hazardous materials such as asbestos, lead-based paints and toxins in ash deposits. If there is any reason to suspect the presence of such materials, (i.e.; Chimneys constructed pre-1975) and there is a risk of exposure of the inspection personnel to those materials, then sampling and analysis of the materials must performed in accordance with the appropriate Occupational Safety and Health Administration (OSHA) and/or National Institute for Occupational Safety and Health (NIOSH) regulations before inspection personnel are permitted to access the chimney.

1.2.2 Class I Inspection

Class I inspections are routine inspections that should be performed at regular intervals of between 6 and 24 months. A Class I inspection is primarily a visual inspection. The exterior of the chimney should be visually inspected, using binoculars or spotter scope from the ground and any nearby vantage points such as building roofs, catwalks, and adjacent chimney ladders and platforms.

While a basic Class I inspection can be performed with the chimney on-line, it is preferable, wherever possible, to perform the inspection with the chimney off-line, to permit access for a full-height inspection of the liner interior, and as much of the height of the annular space as is accessible.

A close-up visual inspection should be made of all parts of the chimney directly accessible by ladders, platforms, or other permanent access to the chimney shell.

1.2.3 Class II Inspections

Class II inspections should be performed at regular intervals in the range 2 to 5 years, and alternated with Class I inspections.

A Class II inspection will be performed with the chimney off-line, and will include all of the work performed for the Class I inspection, plus a full-height interior inspection. Depending on chimney condition and type, a Class II inspection may also include:

a) Full or partial thickness core samples of concrete or brick for laboratory assessment of acid attack, material degradation, remaining strength and corrosion susceptibility.

b) Nondestructive assessment of thickness and weld condition on steel chimneys and flue liners.

c) Nondestructive assessment of FRP liners.

d) Installing crack monitoring equipment on concrete or brick chimneys.

e) Full-height inspection of chimney exterior, particularly to provide direct access to normally inaccessible portions of the chimney, using multiple drops as necessary.

If the inspection is to provide baseline data on which to build a long term monitoring program, items a) through e) above should be mandatory.

1.2.4 Class III Inspections

A Class III inspection is not routine. It is performed only when significant degradation of any structural component has occurred, as indicated by the results of Class I or II inspections, or when the chimney has experienced an unusual event such as:

- Earthquake
- Hurricane/tornado strength winds
- Explosion, Implosion, or significant impact
- Fire or Overheating
- Flooding through failure or leakage of wet precipitation or scrubber systems
- Local modifications that may change performance and applied loads, such as changed air flow and vortex shedding by a new neighboring chimney.

After such an event, a Class I inspection should be performed immediately to determine the need for additional inspection and repair, or to clear the chimney for continued use. A Class III inspection would include:

- Inspection of all parts of a chimney available for a Class II inspection, depending on whether the chimney is on or off line.
- Nondestructive integrity testing.
- Removal of cored or drilled samples of material for physical and chemical analysis.

A Class III inspection should also be performed before any significant structural modifications to a chimney or liner, such as increasing height or installing additional platforms or breechings.

1.2.5 Monitoring Recommendations for Chimneys of Special Importance

It is recommended that especially important chimneys be monitored using more sophisticated techniques than required in Class I and II inspections. The Owner must decide the relative importance of any chimney, but typical cases of especially important chimneys would include large multi-flued chimneys serving several generating units, process plant chimneys serving an entire plant or refinery, vent stacks serving nuclear plants, and similar critical installations where the chimney is often not taken off-line for a period of several years.

Ideally these types of chimneys would be inspected and tested when first constructed, prior to initial operation, to establish a number of structural parameters or 'signatures' which can be measured reliably, with high repeatability. The information from this initial, or 'baseline' inspection should be used to establish a database for the structure.

The data from subsequent inspections can then be compared with the database to determine relative rates of corrosion or other degradation, which is vital in predicting, prioritizing and planning maintenance needs, as well as determining the appropriate schedule and class for each subsequent inspection.

1.2.6 Other Possible Indicators of Special Inspection Needs

While all chimneys need periodic inspection, there are some combinations of structure, appurtenances and operating conditions that should alert the owner to the possibility of higher-than-normal risk of early degradation. Some of the more common combinations are:

- Concrete chimneys with corbel-supported brick liners (no access to the annular spaces for inspection or clean-out of flyash).
- Chimneys with flue gas scrubbers ("wet" systems).
- Lined chimneys with flue gas booster fans (positive pressure gas flow).
- Adjacent chimneys of different heights (flue gas impingement onto the exterior of the taller chimney).

It should also be noted that the newer flue gas emissions reduction systems have little or no track record regarding their possible effects on chimney durability. When relatively new or untried emissions reduction technologies are proposed, the chimney condition should be carefully documented before they are put into effect and closely monitored afterwards, until the owner is satisfied that the technology is having no adverse effect on the chimney.

1.3 Inspection Intervals

The actual interval between inspections, and the appropriate class of inspection will be determined by the chimney type, exhaust gas conditions (composition, temperature and humidity, etc.) fuel, operating conditions and the condition of the chimney and liner system. It follows then, that the deliverable from any Class I, II or III inspection should be a written report that will include recommendations as to the next appropriate inspection type and schedule. In most cases it should be possible for the inspection engineer to provide such recommendations to cover the next five years or so as a maintenance planning aid for the owner.

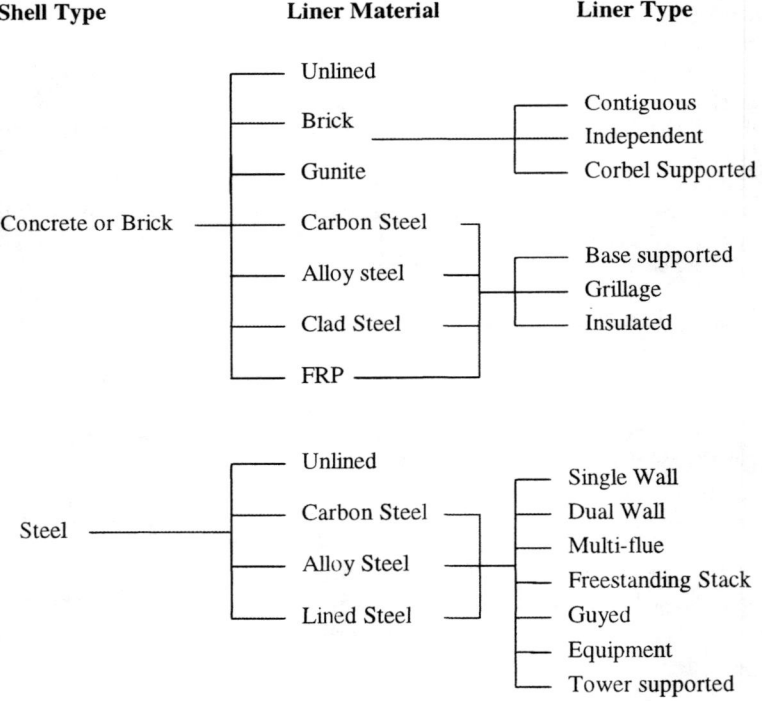

Figure 1: Typical Industrial Chimney Arrangements

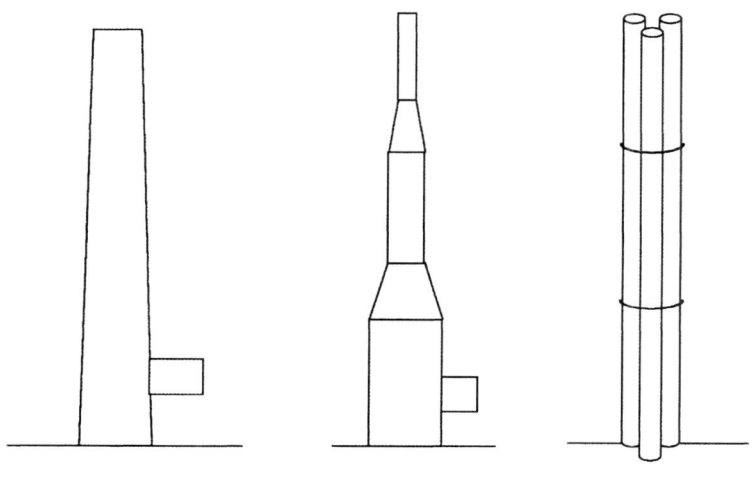

Fig. 2b – Free Standing Brick Chimney

Fig. 2b – Free Standing Steel Stack

Fig. 2c – Clustered Stacks

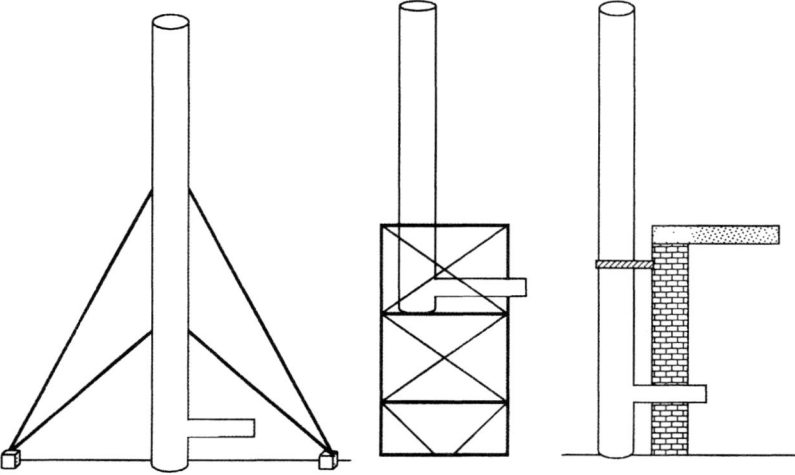

Fig. 2d - Guyed Stack

Fig. 2e – Stack on Flexible Support

Fig. 3f – Braced Stack

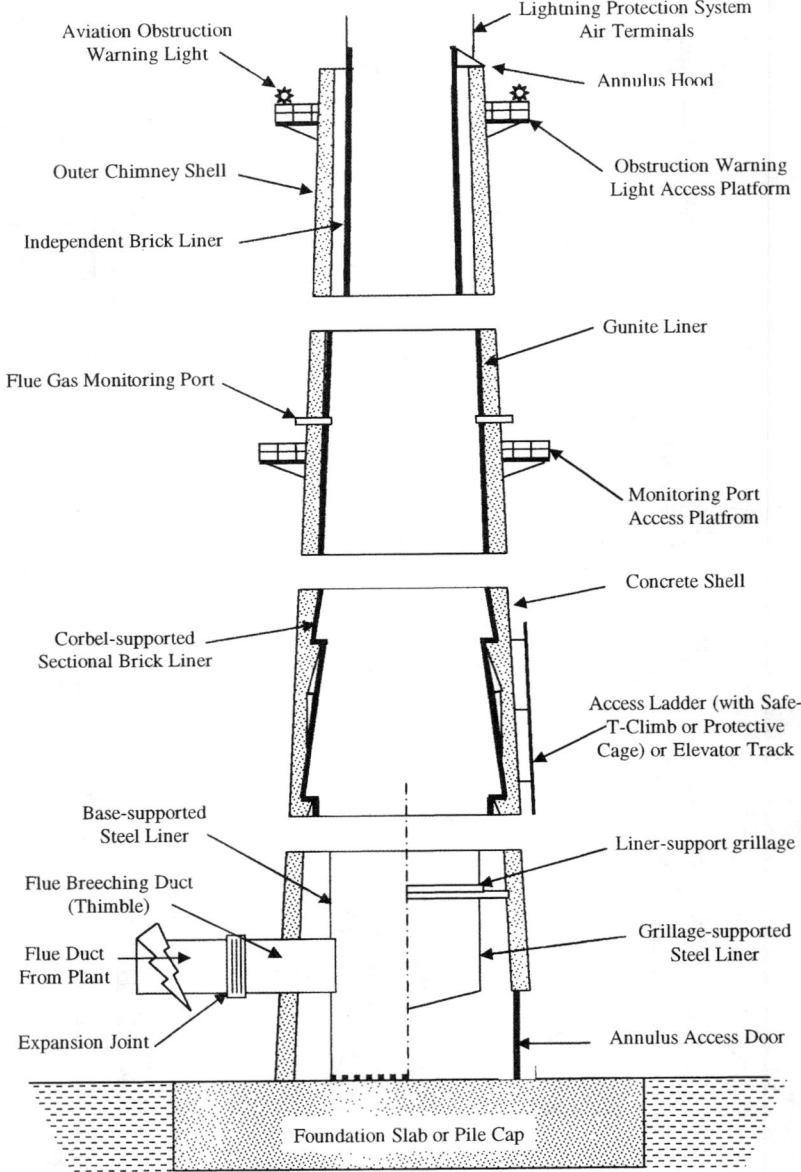

Figure 3: Common Shell/Liner Combinations and Appurtenances

1.4 Table 1: Common Forms of Degradation

COMPONENT	COMMON FORM OF DEGRADATION	PROBABLE CAUSE
CONCRETE SHELL	Cracking, spalling	External stresses, thermal stresses, corrosion of rebar
	Chemical attack	Carbonation triggering corrosion of rebar. Sulfate or acid attack - surface softening and erosion
STEEL STACKS	Corrosion, loss of thickness	Atmospheric & Flue gas corrosion
	Excessive deflection	Underdesigned for wind load
	Excessive cross-wind motion (Fatigue)	Inadequate design for vortex shedding
INDEPENDENT BRICK LINER	Cracking	Structural and thermal stresses, stress concentration at corners
	Mortar and brick softening	Age, acid condensation, changed operation
	Surface Erosion	Age, changed operation, scouring velocity
	Leaning causing problems with hood and interior platform alignment and contact with outer shell	Scrubber operation, salt growth, moisture penetration, over/under or opposite breeching
	Excessive wetness	Very low temperature, (Scrubber) high moisture content
	Degradation of upper courses through rain-triggered acid attack	Normal cooling of gases and wash down, acid condensation
	Positive pressure causing flue gas penetration in airspace	Low temperature (scrubber), loss of draft requiring I.D. fans
CORBEL SUPPORTED BRICK LINER	Acid attack on corbels causing softening of brick and mortar and concrete erosion, rebar rusting, insulation	Penetration of flue gas through cracks in the air space, acids condensing and settling on the corbel

COMPONENT	COMMON FORM OF DEGRADATION	PROBABLE CAUSE
STEEL LINER	Impact damage at bumpers	Improper sliding of the bumpers, too many bumpers causing over stressing
	Corrosion, Loss of thickness	Corrosion of flue gas
	Buckles, ripples (Structural damage)	Possible excessive ΔT, explosion, unexpectedly high temperature excursion
FIBER-REINFORCED PLASTIC (FRP) LINER OR STACK	Cracking, wicking or blistering	Fatigue, overheating, or corrosion
GUNITE LINING	Spalling, cracking, erosion	Age, service related wear and tear, improper material selection
LIGHTNING PROTECTION SYSTEM (LPS) POINTS AND CABLES	Corrosion, broken anchors, broken conn.	Age, ice load, movement of chimney relative to liner with a tight jumper cable, broken during other stack work
	Deteriorated lightning points, Air terminals & ground path	Repeated lightning strikes
HOOD/CAP	Corrosion, structural damage	Age, acid condensation due to cool down effect causing corrosion, excessive movement or lean of the liner/column
AVIATION WARNING LIGHTS (AWL)	Burned-out bulbs Corroded conduit and fittings Damaged or corroded wiring	Age, broken during other stack work
PLATFORMS	Corrosion at joints member, corrosion of grating	Atmosphere corrosion, age, some wash down effects.
LINING BAND	Corrosion, breakage, shearing of bolts	Scrubber operation causing lower temperature, requiring positive draft, infiltration of flue gas and acid attack
MONITORING PORTS	Breakage, spalling of concrete wall or brick, deformation.	Improper design, inadequate room for movement
	Corrosion	Improper material selection

COMPONENT	COMMON FORM OF DEGRADATION	PROBABLE CAUSE
MONITORING PLATFORM	Atmospheric corrosion, overloading	Age, environment, enclosure load
BREECHING LINTEL	Corrosion, deflection	Acid attack from flue gas, inappropriate material for the operation, buildup of flyash on top of flue in annular space
BREECHING DUCT PACKING	Blowout, loss	Vibration, thermal movement, incorrect material, improper installation
BUCKSTAY	Bending out of vertical axis	Inadequate to withstand band tension in the weak axis
LADDER	Corrosion, separation of connection lugs, poor anchorage	Age, use, rigid connection with no provision for vertical stack expansion and contraction
BUMPER ASSEMBLY	Bent members, weld failures, anchor pull out	Inadequate room for thermal expansion and contraction
MULTIPLE BUMPERS OR STAYRODS	Bent members, anchor pull out, weld failure	Excessive loading on lower bumpers
LINER SUPPORT GRILLAGE	Over stressing, noticeable deflection	Fly-ash buildup, inadequate design
EXPANSION JOINTS	Torn, worn out, ineffective	Age, service related wear and tear, some debris deposit
ACCESS DOORS	Corrosion, particularly of hinges and latches.	Service related wear and tear

2.0 GENERAL INSPECTION METHODS AND PROCEDURES

For the sake of brevity, the term 'chimney' will be used throughout this section to mean any exhaust gas emission structure, whether constructed of concrete, FRP, masonry or steel.

2.1 Rationale Behind Inspection Procedures

Visual inspection procedures are inherently subjective. One person may assess the degree of discoloration or a roughening of a surface differently to another observer. Similarly, photographs of the same part of a structure can appear substantially different if taken under different ambient light conditions.

Quantification of observations is therefore important, and the use of nondestructive tests, measurements, and laboratory assessment of material samples can greatly enhance the objectivity of an inspection and subsequent condition assessment. However, changes in ambient conditions can affect the results of some nondestructive tests, by altering the moisture content of concrete or masonry, or the temperature of the surface.

If an inspection is to be performed to establish a maintenance database, or to compare data with that from previous inspections already stored in a database, it is essential that the format of the new data is compatible with the data already stored, and that the inspection methods provide repeatable data. This requires that ambient site conditions, test procedures, and test locations are adequately documented for a third party to be able to accurately re-locate test positions, replicate test procedures, and ensure, where appropriate, that site conditions are replicated, or that due allowance is made for differences in ambient conditions.

2.2 Site Planning an Inspection

2.2.1 General Notes

Due to the typically limited access conditions and large size of most chimneys, the full scope of work required for an accurate condition assessment may not become apparent until the inspection has started. Provisions should therefore be made in bid and contract documents to alter the priorities or the scope of work as dictated by new information obtained once the inspection has started.

Chimney inspectors have many sophisticated tools available today to assist them in quantifying material properties and degrees of degradation. Knowing when and how to use these tools can enhance the quality and quantity of information obtained during an inspection. However, photographing, mapping, or otherwise recording structural defects is meaningless to the owner unless accompanied by a thorough explanation as to the most probable cause of the defects, an assessment of their impact on the overall structural integrity of the chimney, and recommendations for repairs/renovations to correct or prevent the problem from recurring.

Continuity is an important aspect of any ongoing inspection program. Using the same inspector or inspection team each time will not only reduce the effect of any subjective judgments, but will aid greatly in the site planning process since that individual or team will already be familiar with the chimney, access requirements, and general conditions.

2.2.2 Class III and Baseline Inspections

In general terms, a minimum of three equally spaced inspection drop locations will be needed to provide adequate coverage of the exterior of a typical chimney up to about 400 feet in height. Taller chimneys will have larger base diameters and may require additional full or partial height drops to provide adequate coverage of the lower portions of the chimney.

If the chimney is off-line, and access is available to the interior of the liner, at least one full-height drop should be made inside the liner, with particular attention to the flue breeching and ductwork. If noticeable deposits of fly ash have adhered to the liner interior, the liner surface should be cleaned prior to performing the inspection. The choice of cleaning method will depend on the liner type and the quantity of ash build-up, but the purpose of the cleaning is to expose the material of the entire liner surface for direct visual inspection.

If there is an annular space between the chimney shell and the liner, provision should be made for visual inspection of the inner surface of the shell and the outer surface of the liner.

2.2.3 First-time (Baseline) Inspections

If a condition database is to be set up for a chimney where no suitable previous data exists, even though previous inspections have been performed, the new survey should be regarded as a first-time inspection.

The selection of exterior drop positions for a first-time survey should be based on:
- Review of construction documents if available.
- Environmental factors such as prevailing wind direction and neighboring chimneys or cooling towers.
- Accessibility.
- The findings of a preliminary visual survey of the chimney (e.g. observed concentration of degradation).

The preliminary visual survey should include viewing as much of the chimney as possible from grade level, from any fixed access ladders, and from any permanent access platforms. If no permanent platforms are installed on the chimney, a visual survey of the upper portions of the chimney should be performed from grade level using binoculars or a spotter scope. If practical, a temporary platform should be installed at the top of the chimney to provide full circumference access to the top lip of the shell and liner for inspection of caps or rain shields, and lightning protection systems. The temporary platform will also be helpful in placing rigging cables at the desired drop locations.

If the interior of the chimney is accessible, interior drop locations should be selected to provide access to the flue breeching and any areas of visible deformation of the liner. If there are two breechings, then additional partial or full-height drops may be necessary to provide access to the breechings and ducts.

Exterior drop locations should be selected to provide access to any anomalous areas noted during the visual survey. Typical anomalies in a concrete shell would be:

- Bugholed or honeycombed concrete (original construction defects).
- Concrete cracks
- Cracked or peeling paint or coating
- Efflorescence or surface deposits
- Roughened or spalled concrete
- Stains / discoloration (patches, vertical streaks of corrosion products, etc.)

Locations for specific tests on each drop should be selected to include coverage of visible anomalies. If there are no visible anomalies on a particular drop, then physical tests and visual assessments should be performed at regularly spaced intervals. Testing of apparently sound areas is important because the comparison between areas with visible anomalies and apparently sound areas can provide important information as to the cause of the anomaly and the rate of degradation.

The test intervals should be selected to ensure that the zone around each construction joint in the concrete shell is inspected. Where corbel supported liners are installed, particular emphasis should be placed on the zones immediately above each corbel shelf.

Where there is an accessible annular space, inspection procedures will be determined by the size of the annulus, and any evidence of liner leakage or distress such as large deposits of fly ash, or broken reinforcing bands.

If the interior of the chimney is accessible, interior drop locations should be selected to provide access to the flue breeching and any areas of visible deformation of the liner. If there are multiple breechings, then additional partial or full-height drops may be necessary to provide access to the breechings and ducts.

Where there is an accessible annular space, inspection procedures will be determined by the size of the annulus, and any evidence of liner leakage or distress such as large deposits of fly ash, or broken reinforcing bands.

In all cases, the drop positions and test locations or intervals should be documented in such a manner that personnel performing subsequent inspections can easily and accurately re-locate test positions.

2.2.4 Follow-up Surveys

Where the inspection is to be part of a follow-up or condition assessment survey based on a prior database, the initial drop positions should be the same as those documented in the database, and the test intervals should be, at a minimum, also as documented in the database. Additional testing, and additional drops if necessary, should be performed at the location of any visible or suspected anomalies that were not documented in the previous surveys from which the database was compiled.

2.3 Visual Inspection, Photography, and Documentation

Written notes and color photographs should document visible anomalies such as those listed in 2.2.1. Notes for photographs should include observation of weather conditions, time of day, and film speed.

Photographs of anomalous areas, deteriorated fixtures, and visible evidence of distress to the structure or appurtenances should be taken from sufficient different angles of view or elevations to allow full assessment of the visible extent of the damaged area by a third party viewing the photographs.

Where obvious indicators of size or scale such as bricks or ladders are not included in the field of view, some other indicator of scale, such as a ruler, tape measure, or field notebook, should be held directly against the surface being photographed.

Where there is an annular space between the chimney shell and the liner, the inner surface of the shell and the outer surface of the liner should be inspected. Where the annular space is too small for personnel access, a remote video camera can be used to perform the visual inspection. In particular, if broken sections of liner reinforcing bands are visible in the annular space, the designed location for every band should be checked to determine the locations of displaced or missing bands.

Some companies now offer a "Hot Camera" service, where an insulated video camera can be lowered down inside an operating chimney while it is still on line. If it is impossible to schedule an outage at the time that the inspection is required, then Hot Camera inspection may be a useful option. However, flue gases vary in opacity, and the video image from a Hot Camera will rarely have the clarity or detail that direct observation provides. Turbulence in the gas stream will also affect the stability of the image. Hot Camera observation is thus most likely to find large deficiencies such as missing brickwork, and partially collapsed or severely buckled liners, but is unlikely to record narrow cracks or shallow spalling.

2.4 Measurement and Instrumentation

If survey data is to be compiled in a database for comparison with data from subsequent surveys, it is essential that test procedures and measurements are repeatable.

Crack widths should be measured, and the location of the measurement should be documented in terms of orientation, estimated depth, and elevation, or distance from the nearest suitable fixed reference point such as a construction joint, penetration, or fixing point.

Similarly, the location of any other measurements or tests should be documented with reference to fixed points on the structure so that personnel performing subsequent inspections can easily and accurately re-locate those positions.

2.5 Nondestructive Testing

There are several nondestructive test (NDT) and monitoring methods that can provide useful information on the condition of the concrete, masonry, or metal elements of a chimney. If the methods are properly understood and correctly applied, the data from them is highly repeatable and easily incorporated into a condition database. There are, however, a number of misconceptions about some NDT methods, and limitations to their performance, that must be considered when specifying them.

The following list gives brief descriptions of each of the methods. The American Association of State Highway Officials (AASHTO), the American Concrete Institute (ACI), and the American Society for Testing and Materials (ASTM) have produced either standards or codes of practice for many of these methods which are internationally recognized. Where an appropriate standard has been published for the method, the source, reference number and full name is given to aid the reader who wishes to find a more comprehensive description or guidance in the performance of the method. The addresses of the publishing organizations are given in the appendices to this manual.

2.5.1 Carbonation Depth

Fresh concrete is alkaline, with a pH of about 13.5. This alkalinity is an important corrosion inhibitor that reacts with the embedded reinforcing steel to form a passive, corrosion resistant coating. From the moment that the new concrete is exposed to the atmosphere, acid gases in the air combine with moisture to form dilute acids, primarily carbonic acid from carbon dioxide, that react with the alkaline constituents of the cement paste, gradually reducing the pH of the concrete. This process is known as carbonation. When the pH of the surrounding concrete is reduced to less than about 11.5 it is no longer alkaline enough to maintain the passive coating on the reinforcing steel and corrosion will begin if enough oxygen and moisture are present at the steel surface.

Since the carbonation process is a progressive one, originating at the exposed surface of the concrete, monitoring the depth of carbonation and the rate of its penetration into the concrete is useful in predicting when it is likely to reach the reinforcing steel and increase the risk of corrosion. Service life prediction models are also available to aid in this assessment.

An accurate determination of carbonation depth can only be made from petrographic examination of a cored or sawn sample of the concrete, but there is a simple field test that can give a very useful approximation of carbonation depth. If no broken edges are available, the concrete is prepared by drilling a small diameter (13mm) hole. The edge of the hole, or the edge of a break, if available, is flaked off with a small chisel to create a freshly broken surface. A liquid chemical indicator such as Phenol Phthalein, Thymol Phthalein, or *"Rainbow Indicator"* is sprayed on the freshly broken surface. The indicator changes color according to the alkalinity (pH) of the concrete.

A freshly broken surface must be used because surface carbonation of broken concrete occurs rapidly, and drilling crushes the cement paste, mixing material from uncarbonated pores in the cement paste with carbonated material, leading to errors in depth measurement.

Care should be used in interpreting carbonation depth from this test, since it generally only indicates the depth of the fully carbonated concrete. The partial carbonation front, and other atmospheric contaminants, may be deeper than the depth indicated by the chemical test. Multiple samples around the perimeter and at various heights may provide more information concerning the uniformity of carbonation depth and rate of advance, since these are affected by atmospheric exposure (prevailing winds, neighboring chimneys, etc.) as well as concrete condition.

2.5.2 Cover Concrete Depth

If the likely onset of corrosion is to be predicted from carbonation depth measurements, the depth of cover concrete over the reinforcing steel must be known. There are a number of proprietary instruments, or 'covermeters', available for estimating the depth of cover concrete nondestructively. The term "estimating" is used because the type of the reinforcing steel, magnetic or metallic aggregates, and the proximity of neighboring bars or other embedded metal items such as conduits or pipes can affect the accuracy of the depth measurements from these instruments.

Strong electrical or magnetic fields will also affect the accuracy of the measurements, and in extreme cases may make the instruments so unstable that no useful information can be obtained.

For maximum accuracy, covermeter measurements for a particular structure should be calibrated against physical measurement of reinforcing steel depth and size by exposing the steel at a sample location, either by drilling or cutting.

2.5.3 Rebound or Schmidt HammerTest

ASTM C805 - Rebound Number of Hardened Concrete

Also known as the Swiss Hammer, or Schmidt Hammer. A steel impactor is driven against the concrete surface by a spring mechanism, and the rebound of the impactor from the surface is measured, and displayed as a rebound number. In theory the rebound number for a given concrete mix can be correlated with strength, and the Rebound test is generally known as an NDT method for estimating concrete strength. It is also often specified for detecting variations in concrete quality.

In reality, there are a number of factors that affect the rebound of the impactor, and can result in significant variation of the rebound numbers. These factors are:
- Surface coatings
- Carbonation (causes surface hardening of the concrete)
- Surface profile (curved or roughened)
- Surface moisture
- Nature of material at impact point (aggregate or cement paste)

These factors make the Rebound test unreliable for estimating concrete strength on older structures. It can be used to detect zones of anomalous concrete if multiple tests are performed over a given area, but great care should be taken in interpreting the test results. The Rebound test should not be used on painted or coated concrete.

2.5.4 Windsor Probe

ASTM C803 - Penetration Resistance of Hardened Concrete.

The Penetration Test for Hardened Concrete is also known as the Windsor Probe Test. Specially dimensioned and hardened probes are propelled against the concrete surface by a controlled force, typically an explosive cartridge. The depth by which the projectile penetrates the concrete is used to estimate concrete strength.

Test results will vary with concrete strength, but other significant variations will occur for the same reasons as those given for the Schmidt hammer. Multiple tests are required to minimize the effects of these variables, but the penetration of the projectile causes spalling or scarring of the concrete surface. The Windsor probe should not be used on painted or coated concrete or on structures where surface finish is important for aesthetic reasons.

2.5.5 Ultrasonic Pulse Velocity

ASTM C597 – Standard Test Method for Pulse Velocity Through Concrete.

The velocity of an ultrasonic pulse through concrete is a function of the modulus and density of the concrete. Ultrasonic Pulse Velocity (UPV) measurement is therefore useful in determining the uniformity of concrete quality, and locating areas of lower density material caused by distress such as delamination, microcracking or alkali aggregate reaction, or construction defects such as honeycombed concrete.

However, concrete has an inherent variability that must be considered when applying UPV tests. The UPV calculated for a given test will be the average velocity for each particle of cement and aggregate along the pulse transmission path. If the UPV transmitter or receiver are moved an inch or so across the surface, the pulse may pass through a different set of particles. Aggregate size, orientation, and distribution may differ, resulting in a higher or lower average velocity. If the UPV test is to be used to monitor degradation by repeating tests over a period of time, it is essential that the actual locations of the transducers be accurately marked or documented, and precisely repeated for each set of tests. Experience has shown that variations in transducer locations can result in apparent velocity variations that render the data useless for trend tracking.

2.5.6 Half-cell Potential / Resistivity

ASTM C876 – Standard Test Method for Half-Cell Potentials of Uncoated Reinforcing Steel in Concrete.

The most common form of corrosion on reinforcing steel is oxidation of the steel. Oxidation is an electrochemical process that requires oxygen, water, and a circuit through which electrical current can flow. Dry concrete normally has a high resistivity that impedes the flow of corrosion currents. Moisture or contaminants in the concrete lower its resistivity, and can make it sufficiently conductive that it forms one part of the circuit, and the reinforcing steel forms the other part.

The energy required to produce the electrical current comes from differences in the electrical potential of the reinforcing armature. These potential differences can result from surface contaminants on the steel itself, and stresses caused by the original production and bending of the steel, and subsequent structural stresses. Higher potential differences and lower concrete resistivity increase the likelihood of corrosion.

The half-cell potential method is a means of measuring the potential difference between various points on the reinforcing steel, and so determining where or when corrosion activity is likely to occur. An electrical half-cell, most commonly made of copper in copper sulfate ($Cu/CuSO_4$), or silver in silver chloride ($Ag/AgCl$), is connected in series with a sensitive voltmeter, and a reference location on the reinforcing steel. Pressing the tip of the half-cell against the concrete surface over another part of the reinforcing steel completes a measurement circuit. The potential difference (PD) between the test point and the reference point is then measured and recorded.

A fundamental weakness of the method is that the conductivity of the concrete will vary as moisture content changes, so changing the apparent PD value measured. Similarly, electrical discontinuities in the reinforcing armature, or poor contact between the half-cell and the concrete surface, will affect the test results. Interpretation of test results can be made more reliable if PD data is correlated with measurements of the concrete resistivity taken at the same time. Poor correlations between PD and resistivity values indicate invalid test results due to one or all of the factors given above.

Since ambient conditions such as humidity and temperature will affect the measured values, PD/resistivity data should not be interpreted in terms of absolute values. More reliable estimates of the probability of corrosion activity are made from assessing the difference, or gradient, between adjacent parts of the structure.

Any chlorides that may be present on the surface of the concrete also adversely affect the accuracy of a copper/copper sulfate (Cu/CuSO4) half-cell. Since chlorides are present in abundance in the flue gas from most coal burning plants, investigation of concrete chloride content should be performed by laboratory analysis of drilled powder samples before using Cu/CuSO$_4$ Half-cells to determine potential difference.

Another consideration is that the test is not truly nondestructive. If spalling of the concrete has not already exposed reinforcing steel, access for the reference connection to the reinforcing must be provided by drilling or chiseling a small hole in the concrete. The half-cell potential test cannot be used on reinforcing steel coated with epoxy or any other non-conductive material.

Access is another consideration. The test is most effective and the data most easily interpreted if large surface areas can be surveyed and mapped. On chimneys, however, the limited access afforded by ladders, platforms and swing-stages can make it difficult to apply the test to large contiguous areas.

2.5.7 Impulse Response Test

The Impulse Response Spectrum test is also sometimes known as the Sonic Mobility test, or the Transient Dynamic Response (TDR) test. The surface of the structure is instrumented with a geophone velocity transducer. The structure is struck with a small sledgehammer, which contains a load cell to measure force input. The data from the two instruments is recorded and processed on a PC-based data acquisition system, where both signals are converted into frequency-based data, and velocity is divided by force to provide a normalized response of mobility against frequency, or the Impulse Response Spectrum (IRS).

The IRS contains information on the dynamic stiffness of the test point and the quality of the concrete. The integrity of the structure at the test point can be assessed from the IRS data, and the test can be used to identify conditions that are not visible to the eye, such as sub-surface honeycombed concrete, delamination, micro-cracking, and lower modulus concrete. The dynamic stiffness value is particularly useful for identifying anomalous zones on uniform or symmetrical structures such as chimneys and silos.

The IRS test is unaffected by surface moisture or paint, but thicker elastomeric coatings can deform at the impact point, leading to variability of stiffness measurements.

The IRS test measures minute vibrations of the surface of the structure. The test results may, therefore, be difficult to interpret if the structure is vibrating heavily, such as a chimney with a poorly balanced booster-fan or one with turbulent flue gas flow.

2.5.8 Impact-Echo Test

ASTM C1383 – Standard Test Method for Measuring the P-Wave Speed and the Thickness of Concrete

The impact-echo test is similar in principle to the IRS test, but a much smaller ball-bearing impactor is used, and the response of the surface is measured with either a displacement transducer or an accelerometer. The smaller impactor produces a higher frequency input than the IRS test, but much lower energy. The test therefore does not provide dynamic stiffness or mobility data, but can be used to measure thickness with a relatively high degree of accuracy.

Due to the size of the impactor and the limited range of the displacement transducer, the impact-echo method is not suitable for use on roughened or spalled surfaces. Where surface roughness is excessive, grinding of the concrete is required to provide a suitable test surface. Thick elastomeric coatings also hamper the method.

2.5.9 Crack Monitoring

Where significant cracks are observed, monitoring of crack growth and/or movement can provide valuable information for condition assessment and durability prediction, and in some cases can help determine the cause of cracking.

If accurate long-term monitoring of crack movement is considered to be important for a particular structure, the crack should be instrumented with a suitable indicator such as the Avongard crack monitor, or with locator studs for a mechanical displacement gage such as the 'Whittemore' or 'Demec' gages.

For normally inaccessible areas, crack movement monitoring can be automated by installing resistance or vibrating wire strain gages and a suitable recording system. A multiple-gage arrangement can be used to monitor crack-width changes and longitudinal movement. However, these gages can accommodate only a limited range of movement, and care should be exercised in selecting the gage type to ensure that the working range of the gage selected for the project is large enough to accommodate the anticipated crack movements.

2.5.10 Ultrasonic Thickness Gage

ASTM E797 - Practice for Measuring Thickness by Manual Ultrasonic Pulse-Echo Contact Method

Steel chimneys, steel liners and steel flue duct systems can lose cross-section due to erosion, corrosion, or both. Access to one side of the steel is needed in order to measure the remaining cross-section or residual thickness using an Ultrasonic Thickness Gage.

The principle of the Ultrasonic Thickness Gage is the same as for the UPV test described earlier, except that the system uses higher frequency pulses and smaller transducers. The transmitter and receiver transducers are mounted side-by-side in a common sensing head, which is placed on the accessible surface of the metal to be examined. The emitted pulse propagates through the material to the opposite face, where most of the energy is reflected back to the receiver. The round-trip time for the reflected pulse is measured. If the UPV for the material is known and entered into the control unit, then thickness can be automatically calculated and displayed.

A gel-type acoustic couplant is normally required to ensure good acoustic transmission between the transducers and the metal surface, where minor surface irregularities may exist. Typical products are petroleum-based gels such as Vaseline, or Water/Cellulose gels similar in consistency to KY jelly or wallpaper paste. If surface corrosion or contamination have resulted in a roughened or scaly surface, it will usually be necessary to clean the surface back to bare metal by grinding, grit-blasting, or similar means. Thick coatings such as elastomeric sealants or buildup from multiple coats of paint should also be removed prior to performing this test.

2.5.11 Ultrasonic Weld Inspection/Crack detection

ASTM E164 – Standard Practice for Ultrasonic Contact Examination of Weldments

Ultrasonic Testing for weld inspection and crack detection in metals is commonly known as UT. The principle of the equipment is similar to the Ultrasonic Thickness Gage, but the control unit includes a display screen and a set of analysis tools for exhibiting and measuring the actual received waveform.

Analysis of UT data can reveal flaws or cracks in welded metal joints, and can detect fine cracks and delaminations in metal structures. Acoustic couplant, surface condition and preparation requirements are the same as for the Ultrasonic Thickness Gage.

2.5.12 Thermal Imaging

An Infra-Red (IR) camera produces images using apparent temperature differences instead of light and dark, as a black and white film does. Digital IR images can be colorized to make subtle temperature differences easier to identify.

The term 'apparent temperature' is used because the actual property measured is the IR emissivity of the surface being imaged. The IR emissivity is determined by a combination of the temperature of the surface, the material of the surface, the moisture content, and the amount of IR impinging on the surface from other sources, such as solar radiation or nearby heat sources.

Thermal imaging is widely used to monitor and locate hot spots in electrical switch and breaker panels, motors and transmission lines. Thermal imaging also has a proven track record in monitoring energy efficiency by detecting insulation failure in building envelopes. In chimney inspection and maintenance, thermal imaging is useful for identifying insulation loss or failure around boilers and flue ducts, and detecting debonding or loss of refractory linings in ducts and pipes.

Thermal imaging can also be used to identify temperature anomalies on the exterior of chimney shells. However, care should be taken when interpreting IR images. Temperature anomalies may be caused by insulation loss, flue gas leaking through the liner, turbulent airflow around the chimney, or, in the case of brick or concrete shells, variations in moisture content. Under certain circumstances when the ambient temperature is changing relatively quickly, such as at dawn or dusk, delaminated or cracked concrete can be detected in IR images because the smaller mass of the delaminated section warms or cools faster than the main body of the chimney.

Accurate interpretation of IR images in many cases will require additional investigation to establish the cause of the temperature variation.

2.5.13 Subsurface Interface Radar

Subsurface Interface Radar (SIR) is commonly used for geophysical surveys, and so is also known as Ground Penetrating Radar (GPR). For this method a central processor unit controls the emission of a series of very high frequency electromagnetic pulses from a transmitting antenna, and measures the time taken for any reflections from each pulse to be detected by a receiving antenna.

The antennae used for examination of concrete and masonry structures are contained in a single unit, which is moved slowly over the surface of the area being evaluated as the pulses are emitted. Where the pulse encounters an interface between materials with differing dielectric properties, a portion of the pulse energy is reflected back toward the surface. Typically, in a chimney, at an interface with a high dielectric difference, such as steel reinforcing embedded in relatively dry concrete, or steel reinforcing bands around the outside of a brick liner, a high proportion of the energy will be reflected, and a strong signal will be detected.

Typically the radar data is displayed as a vertical chart, with time in nanoseconds as the Y-axis. As the antennae pass over the surface and each subsequent measurement is made the display is moved forward on the screen a fraction of an inch, and the new signal is printed beside the preceding one. Thus, a series of measurements are shown graphically as a vertical 'slice' through the material over which the antennae passed. The horizontal axis represents distance, scaled according to the speed at which the antennae were moved over the surface.

SIR is useful in identifying the location of liner reinforcing bands where there is no access to the exterior of the liner, and for mapping reinforcing steel where the embedment depth is beyond the range of electromagnetic covermeters.

2.6 Sampling and Laboratory Testing of Concrete and Masonry

Chimneys are inherently aggressive environments due to elevated operating temperatures and the chemical content of the exhaust gases. If useful predictions of remaining life or maintenance needs are to be made based on the results of an inspection, the causes of any detected degradation must be identified. Since a number of degradation mechanisms in concrete chimneys and masonry stacks involve chemical attack on the concrete or masonry, or chemical change within the concrete or mortar itself, it is often very useful to perform some limited sampling and laboratory analysis.

For some tests powder samples can be taken from concrete or masonry chimneys by drilling small diameter holes. Other tests require a larger sample of material, which is typically obtained by core drilling or sawing.

2.6.1 Concrete Chloride Content Analysis

AASHTO T260 – Standard Method of Sampling and Testing for Total Chloride Ion in Concrete and Concrete Raw Materials

Exhaust gases from fossil fueled plants can contain a high proportion of Hydrochloric Acid (HCl) vapor, which, at normal operating temperature, is discharged with relatively little harm to the flue or chimney system. However, if the temperature of exposed steel or concrete surfaces is below the dew point of HCl, the acid condenses and directly attacks the material. Free chloride ions from HCl penetrate concrete and masonry rapidly, where they act as catalysts in the corrosion of embedded steel such as reinforcing or conduit. Concrete that is heavily contaminated with chlorides cannot be effectively repaired without first removing all the contaminated material.

Chloride content can be determined in a laboratory from a drilled powder sample of concrete or masonry. The rate of chloride penetration and the likely direction from which contamination is occurring can be determined by analyzing powder samples taken from different depths in the same drill hole.

2.6.2 Petrographic Analysis of Concrete or Brick/mortar Masonry Samples

ASTM C856 – Standard Practice for Petrographic Examination of Hardened Concrete

A full petrographic analysis requires a cored or sawn sample of the material. Petrographic analysis of concrete can determine the overall quality, condition and porosity of the material, the true depth of carbonation or other chemical attack, and identify the nature of any such attack. Where corrosion of reinforcing steel is known or suspected to have occurred, the core sample can include a section of the steel, allowing the quality of the bond between the steel and the concrete to be assessed, and, if corrosion has occurred, the residual thickness of the steel can be accurately measured.

The petrographic analysis can also assess air void distribution and permeability to estimate susceptibility to freeze/thaw degradation, and penetration of carbonation or chlorides.

2.7 Sampling and Laboratory Testing of Metals and FRP

For metal or FRP chimneys and liners it may be necessary to cut out small coupons of the material for laboratory testing.

2.7.1 Brittleness - Charpy and Izod Methods for Metals

ASTM E23 - Standard Test Methods for Notched Bar Impact Testing of Metallic Materials

The notched bar impact tests are used to determine the brittleness of metals. A sample coupon of the stack or liner material is removed, shaped and prepared according to the test method selected. It is mounted in an anvil as a simply supported beam for the Charpy Test or as a cantilevered beam for the Izod Test. The sample is subjected to an impact from a simply suspended pendulum that has more than sufficient mass and momentum to fracture the sample in a single impact. The pendulum energy absorbed by the sample prior to fracture and the nature of the fracture provide information about the fracture toughness of the sample.

The Notched Bar Impact Test methods may be performed at various temperatures to aid in predicting the performance of the stack or liner when subjected to sudden loads such as wind gusts or seismic events, under a variety of ambient weather conditions.

2.7.2 Barcol Hardness Test for Resins

ASTM D2583 – Standard Test Method for Indentation Hardness of Rigid Plastics by Means of a Barcol Impressor

The surface hardness of fiber reinforced plastic (FRP) can be checked using the Barcol Impressor Hardness Test. The expected Barcol readings vary for each resin but in general should be between 40 and 50. For maximum accuracy a sample of the material should be removed from the stack and placed on a smooth hard surface. Multiple tests should be performed for each sample, but the actual number of tests required depends on the typical Barcol reading. ASTM D-2583 recommends the number of tests for each range of hardness values.

2.8 Dynamic Monitoring and Modal Analyses

Certain chimneys, due to safety, environmental, and financial reasons, may be deemed more important than others. For these chimneys dynamic monitoring or modal analysis may be desirable after a significant event which may affect structural integrity, such as a sudden advance in observed degradation, or exposure to earthquake, hurricane or explosion. The maximum benefit can only be derived from such monitoring if a pre-event baseline for the structure has been established.

The dynamic behavior of a chimney can be documented by instrumenting the structure and recording its response to excitation. The instrumentation would typically consist of accelerometers and/or velocity transducers mounted at several points over the height of the structure, and in at least two axes, i.e. north-south and east-west.

Additional instrumentation may consist of strain-gages, displacement transducers, and laser surveying stations.

The instrumentation should be connected to a data-acquisition and recording system. For maximum accuracy the data-acquisition system should be capable of recording all instrument channels simultaneously, or in high-speed burst mode. Excitation of the structure can be either from natural forces, or by induced loading. The most common natural force is wind, but this procedure is obviously weather-dependent, and may mean monitoring the structure for an extended period of time before the desired loading or movement is achieved.

Induced loading has been successfully applied by cable tensioning, and by eccentric shaker. For cable tensioning a large cable is affixed at some point near the top of the structure, and put under tension by winch or similar device, thus loading the structure eccentrically. The cable is released suddenly, effectively 'twanging' the structure, and allowing it to return to equilibrium at its natural pace.

An eccentric shaker typically consists of two counter-rotating eccentric masses. The direction of excitation is controlled by the relative phase of the two masses, and varying the speed at which the masses rotate controls the frequency, or period, of excitation. The eccentric shaker system is marginally more expensive to employ than the cable tensioning system, but does not have the potentially significant hazard of a heavy uncontrolled cable recoiling across the site or falling to the ground.

The instrumentation records the behavior of the structure under the selected excitation. Assuming that the testing has been performed on a sound structure, the data will provide a baseline 'signature' for the structure. Comparison of the database with the results of a modal analysis of the structure should show an excellent correlation if the structure is in fact sound, and performing as designed. If it is considered possible that structural integrity is impaired by exposure to an excessive load situation such as earthquake or hurricane, etc., then the dynamic monitoring should be repeated. Comparison of the new data with the original baseline data will reveal any significant differences in behavior caused by loss of structural integrity, and will aid in a risk assessment of the damaged structure.

It should be noted, however, that the sensitivity of this method varies considerably from structure to structure. In some cases considerable degradation of the structure must occur before it becomes apparent in dynamic tests. For this reason it is important that the test parameters and conditions are recorded in great detail to ensure that operators of any subsequent tests can replicate the original conditions as closely as possible.

2.9 Typical Inspection Program Specifications

The actual methods specified for an inspection program will depend on the structural type of the chimney, the type and purpose of the inspection, and/or known conditions or problems. Some recommended combinations of methods for each class of inspection program are given in these guidelines. The owner should establish what is to be accomplished in the desired inspection(s) before the work is contracted or started.

If an owner is unsure about what class of inspection or what selection of methods would be appropriate, consultation with an experienced chimney engineer or inspector is recommended before procurement documents or specifications are finalized.

All inspection records should be retained by the owner, and made available to the inspector if required. It greatly simplifies the process and increases the consistency of report recommendations if inspection format and record-keeping procedures are established at the beginning of a program, and all subsequent inspections are reported and recorded in the same format.

2.10 Other Information to Include in Bid Request Documents

Pre-bid site visits by prospective inspection firms may not always be convenient or practical. If competitive bids for a chimney inspection are to be sought via Request For Proposal (RFP) documents, it is important that each bidder is fully informed about the structure of the chimney in question, and any particular concerns that the owner may have. The information in an RFP package should include:

- As-built drawings of the chimney, if available, or
- A brief description of the shell and liner system construction
- Proposed chimney status at the time of inspection e.g. On-line or off-line
- Access limitations caused by proximity of other structures
- Present appurtenances, such as platforms or elevators added since construction
- Purpose of the inspection and owner's concerns: i.e., visible damage, exposure to excessive loading, etc.
- Whether the chimney liner interior will be cleaned prior to the inspection, and if so, who will be responsible for the cleaning?
- Summary of results of past inspections

2.11 Record Keeping /Report Archives

The appropriate form and duration of record keeping or report archiving will, to some extent, depend on each owner's available filing system space and corporate mandates. The inspection reports document changes in the condition of a chimney over time.

When coupled with maintenance and repair histories, they form a valuable resource for any plant manager charged with the upkeep of a chimney.

If it is not possible to archive all reports in full for the entire life of the chimney, then the salient points in each report should be compiled into a condition-tracking database. At a minimum, any Class II or III inspection report should be held until the next Class II or III inspection, and should be available to the firm performing the inspection.

From the archived reports or the condition database it should be possible to track trends and rates of degradation, as well as identify sudden changes in type or rate of degradation. If these trends or changes can be correlated with specific events such as change of fuel, modifications to the plant or the chimney or major repairs, then the cause of the degradation can usually be more accurately identified, and appropriate preventive or remedial action can be planned.

2.12 Minimum Inspection Team Qualifications and Safety Training

The integrity and value of a chimney inspection, whether Class I, II or III, are greatly affected by the qualifications, capabilities and experience of the inspection team.

2.12.1 Technical Qualifications

At a minimum, the inspection should be supervised and reviewed by a registered professional engineer with a background and training in the design, inspection and/or repair of chimneys and stacks, or a person who has acquired equivalent qualification through experience in all aspects of the inspection and reporting process. When evaluating experience, consider the variety of structures, work and conditions that form that experience.

The qualifications and experience of the rest of the inspection team may vary according to the type of chimney (e.g: concrete versus steel), class of inspection (e.g: Class I - visual versus Class II – detailed), physical condition and safety significance of the structure, accessibility, and data needs the post-inspection structural evaluation. Inspection personnel should possess sufficient knowledge and experience to be able to identify degradation and performance problems. as well as being familiar with visual, nondestructive and destructive testing methods and their application to chimney structures.

2.12.2 Safety Training

Safety cannot be stressed too highly, particularly when working at heights or when other hazardous conditions or materials may be present. The person responsible for supervising the team on site should be qualified as a supervisor by reason of training and experience in fall protection, confined space entry and avoidance of exposure to hazardous materials, according to the requirements of the Occupational Safety and Health Administration (OSHA) regulations in OSHA 29 CFR 1926.

All other members of the site inspection team should be qualified in fall protection, confined space entry and avoidance of exposure to hazardous materials by reason of training to at least the minimum level required by the OSHA regulations quoted above.

In addition to the OSHA requirements, the inspection team should be familiar with, and comply with any plant-specific safety plans, lock-out/tag-out procedures, etc., that apply to the work area.

3.0 CONCRETE SHELLS

3.1 General Information

Chimney shells constructed of concrete are generally reinforced with steel embedded in the concrete, and maybe constructed in almost any cross-sectional shape, such as square or oval, but most are circular in plan. Short, large diameter chimneys may be cylindrical in shape, but most concrete chimneys taper to a narrower cross-section at the top.

A meaningful inspection of a reinforced concrete chimney column requires a thorough understanding of the design, construction, and operation of the chimney since most problems associated with the structure can be linked to one or more of these subjects.

Since the early 1930's, the ACI has developed concrete chimney design standards. Throughout the past 65+ years, many significant design changes have taken place due to recognized design deficiencies, advances and refinements in the prediction of design loads, and changing design philosophy. Some of the major design related issues addressed over the years are as follows:

- Recognition of thermal stresses
- Recognition of radial wind loads
- Potential for reverse thermal gradients
- Two face vs. single face reinforcing steel
- Working stress vs. ultimate strength design approach
- Improved wind load criteria
- Improved earthquake load criteria
- Recognition of across-wind loads
- Potential for the coupling effect of the liner and column

Knowing the standard in effect at the time of original design, and understanding the structural design implications of that standard will assist the individual planning the inspection and alert them to potential problem areas. Furthermore, it will greatly enhance the ability to accurately assess the present structural integrity of the chimney.

3.1.1 Operation:

The operation of the chimney can have a direct impact on the condition and serviceability of the concrete column. Changes in flue gas temperature and flow rates relative to ambient can affect thermal stresses, condensate levels, ventilation requirements, plume rise, and liner movements. Being able to assess the implications of the operating conditions will assist the individual performing the inspection and alert them to potential problem areas.

3.2 Construction Materials

Two basic cast-in-place construction techniques, slipform and jumpform, have been used to erect the majority of concrete chimneys. More recently, several precast concrete chimneys have also been erected. Each construction technique has its own signature traits and potential pitfalls.

The type of chimney lining and its support arrangement can have a significant impact on the behavior of the concrete column. In addition, older chimneys typically may have undergone numerous repairs and possibly significant structural renovations.

Being able to recognize the type of construction used and to comprehend the structural implications of potential lining/column interaction or structural modifications will assist the individual performing the inspection and alert them to potential problem areas.

3.3 Degradation/Modes of Failure

3.3.1 Chemical Attack

In most routine boiler operations, hydrochloric and sulfuric acids are the primary chemical compounds in the flue gas stream responsible for degradation. Wherever the flue gas comes in contact with the concrete, there is likely to be damage. Of particular concern are chimneys fitted with forced draft systems, made necessary by lower flue gas temperature caused by emissions reduction systems, and consequent lower flue gas flow velocity. The forced draft systems create a positive pressure in the flue, which was most probably designed and built as a negative pressure convection draft system. As a result, flue gas leaks through cracks, joints and damaged seals in the liner system to impinge on the inner surface of the shell.

In simple terms, the sulfuric acid will attack various compounds present in the Portland cement (i.e. tricalcium aluminate and calcium hydroxide) and the resultant product forms gypsum. The formation of this by-product is also accompanied by a significant increase in volume (approximately 20%) and a considerable amount of pressure which weakens the bond between the cement paste and the aggregate. The results include popout of aggregate, delamination of the concrete, and a cracked, friable surface that eventually causes loss of cross-section due to erosion. Damage is exacerbated by the presence of moisture.

Free chloride ions released by the hydrochloric acid rapidly penetrate the concrete pore structure, and migrate through the material according to the flow of electrical currents generated by potential differences on the reinforcing armature. The chloride ions become concentrated at anodic points on the reinforcing steel, where they act as catalysts to the corrosion process. In the corrosion process, particles of steel are removed from anodic areas and deposited in cathodic regions.

Where corrosion is widespread, this deposition of material forms a flake-like structure (rust) that occupies up to eight times the volume of the original steel. This increase in volume creates stresses that eventually crack the concrete, and can cause spalling of the cover layer. Typically streaks or stains will be visible on the surface once cracking has been initiated. In the case of chloride-induced corrosion, however, the process is more concentrated, and is often referred to as pitting corrosion. Because the corrosion is concentrated in a small area, the entire cross-section of the reinforcing steel can be corroded away before enough scale has built up to cause cracking and give visible warning of the corrosion damage.

3.3.2 Mechanical and Thermal Stresses

Concrete is strong in compression, but relatively weak in tension, so structures rely on reinforcing steel embedded in the concrete to provide tensile strength. Under normal conditions it is an almost symbiotic relationship, where steel provides the tensile strength for the concrete, and the concrete, in turn, protects the reinforcing steel from corrosion. However, when mechanical or thermal stresses cause cracking of the concrete, the cracks can provide easy access to the steel for atmospheric pollutants, moisture and oxygen, thus initiating the corrosion process.

Chimneys that are subjected to load cycling or on/off duty cycles can experience rapid large-scale changes in temperature, and are therefore especially prone to cracking as a result of thermal stresses. Some portions of the chimney heat up or cool down more quickly than adjacent portions, and the resulting temperature differential causes differences in the degree of expansion or contraction of the various portions which results in cracking and spalling of the concrete. Rapid heating of thick shells when air temperatures are very low will cause a high thermal gradient through the thickness of the concrete, thus generating high stresses.

High mechanical stresses and subsequent damage can also be caused by excessive wind loads, earthquakes, or impacts.

3.3.3 Weathering

Weathering and freeze-thaw cycles are also often responsible for degradation, particularly in colder climates. All concrete is, to some extent, porous. Changes in atmospheric temperature, humidity and pressure result in a movement of air and moisture through the concrete. One result of this is the carbonation process, a chemical reaction that progressively lowers the pH of the cement paste until the concrete can no longer protect the reinforcing steel from corrosion.

In colder climates, if a chimney has a duty cycle that is long enough to allow the concrete to cool to ambient temperature and sufficient moisture to penetrate the surface of the concrete a second result can be scaling of the surface caused by alternate freezing and thawing of moisture that has penetrated the concrete. As the moisture freezes it expands, and causes tiny cracks within the concrete. After a few freeze-thaw cycles the surface concrete becomes so weakened that it begins to flake or peel off. This process can eventually expose the reinforcing steel, and so initiate corrosion.

Freeze-thaw damage can be minimized and sometimes eliminated altogether by using an air-entraining admixture in the concrete during construction. Proper air entrainment creates a system of air voids that provide room for the expansion of ice as it forms without significantly increasing the permeability of the concrete.

3.4 Inspection Programs

All of the foregoing items will influence the content of a properly planned chimney inspection. It is suggested that, before finalizing an inspection program, the reader refers to the following publications by the American Concrete Institute: ACI 201.1R - Guide for Making a Condition Survey of Concrete in Service, ACI 228.2R - Nondestructive Testing of Concrete, ACI 207.3R - Practices for Evaluation of Concrete in Existing Massive Structures for Service Conditions, and ACI 307 - Design and Construction of Reinforced Concrete Chimneys.

3.4.1 Class I Inspection

a) Visual Inspection

The exterior of the concrete shell or concrete liner should be visually inspected from existing access points (i.e. ladders, service platforms, man-lifts, etc.) when available.

This visual inspection should serve to examine and record the condition of the following areas or items closely:

- Condition of Concrete - Cracks, spalled, or honeycombed areas in the structural wall. All visible exposed reinforcing bars should also be noted. Close attention should be paid to areas around existing penetrations or beam pocket openings. Noticeable staining on the exterior or interior surfaces should also be recorded.
- Coatings (Aviation warning paint, surface sealants, etc.) – Areas of deteriorated, peeled, or weathered coatings should be noted.
- Construction or cold joints - Joints should be visually checked for cracks and signs of leakage.
- Lightning protection system. Items such as air terminals, downleads, lead covering on the uppermost 25' of system, anchors, and splice connections should be inspected for discontinuities, fraying, corrosion, erosion and anchorage condition.
- Chimney cap or annulus rainhood – If accessible, the condition of the steel or fiberglass cap or rainhood should be inspected for degradation, broken components, and cracks. The anchorage to the top of the concrete shell or liner should also be inspected.
- Aviation Warning Obstruction Lighting (AWOL) – The condition of each light, light support, and associated wiring should be inspected. All broken lights should be identified.
- Breeching duct – All areas of the exposed breeching duct exterior plate (or lagging) and at the concrete shell or liner penetration should be inspected. The duct should be inspected for missing or unsecured lagging, exposed insulation, and exterior degradation of the duct plate and stiffeners. The gap between the duct (plate or lagging) and the concrete wall should be inspected for evidence of flue gas leakage and rainwater entrance.
- Platforms and ladders – All members, bolted and welded connections, and finish on all materials should be inspected for overall condition. The condition of anchors used to secure members and brackets should also be inspected.
- Access doors – Mandoors and roll-up doors should be inspected for their general condition and operation. Door hinges, latches, and seals (particular in pressurization systems) should be inspected and their condition noted.
- Construction opening door - The condition of the construction opening closure material, doors, frame, hinges and anchorage to the concrete shell should be inspected for degradation and proper operation.
- Exterior tension bands - The quantity, location, and condition of exterior bands should be noted as well as that of the bolted or welded splice connection.

- Exposed areas of concrete foundation or pedestal - Cracks, spalled concrete or exposed rebar should be identified.
- Annulus pressurization system – The condition of the pressurization fan, fan motors, ductwork into the concrete shell and all accessible opening seals should be inspected for holes, tears and gaps.
- Elevator – The general condition of all accessible elevator components should be inspected and noted. Qualified elevator technicians should perform thorough inspections of chimney elevators at regularly scheduled intervals.

3.4.2 Class II Inspection

a) Visual Inspection

The visual inspection of the concrete shell or concrete liner should include the examination of items identified in Section 3.3.1, performed from any existing access points, (i.e. ladders, elevators, service platforms) as well as from a full height exterior transit (drop) approximately 180° from the ladder. The two full-height areas of the chimney (ladder side and 180° from the ladder) should be closely examined to determine the condition of the areas and items listed below. Where accessible, an inspection (or drop) should be made to inspect the interior condition of the concrete shell or concrete liner.

- Cracks – Cracks should be identified and mapped onto a chimney development sketch. The range of elevation, maximum and minimum width, and orientation of each crack should be noted.
- Spalled concrete – Areas exhibiting spalled concrete should be identified and depicted on a chimney development sketch. The elevation, size of spalled area, and orientation of the spalled area should be noted.
- Exposed rebar – Any exposed rebar should be identified and depicted on a chimney development sketch. The elevation, orientation, amount and condition of reinforcing bars and the thickness of the cover concrete adjacent to the exposed bars should be noted.
- Honeycombing or areas with excessive surface voids – Areas exhibiting honeycombing or excessive surface voids should be identified and depicted on a chimney development sketch. The elevation, size and orientation of honeycombed areas should be noted.
- Staining – Discoloring of the concrete due to rusting of metal components (doors, vents, embedded items, etc.) or leakage from liners should be identified and depicted on a chimney development sketch.
- Construction or cold joints – All cracks and visible areas of leakage in construction joints should be identified and depicted on a chimney development sketch. The elevation and orientation of cracked or leaking

areas should be noted. Previous patching or grouting material used to repair the construction joints should also be noted.
- Openings – The condition of the concrete wall local to all openings (construction opening, beam pockets, access openings, port and light doors, etc.) should be inspected and noted. Defects such as cracks, spalling, and local corrosion from embedded items should be identified.
- Coatings – The condition of all exterior coatings (paint, aviation striping, surface sealants, etc.) should be checked for degradation, peeling, or weathering (fading).
- Chimney cap or annulus rainhood – The structural condition of the steel or fiberglass cap or rainhood should be inspected for degradation, broken components, and cracks. Connection welds and bolts should be checked for evidence of degradation. Surface condition (peeled paint, rust, exhaust buildup, etc.) should be noted. Anchorage to the top of the concrete shell or liner should be thoroughly inspected for missing or deteriorated bolts and condition of any embedded items.
- Platforms and ladders – The structural condition of all members and connections (bolts, welds) should be inspected and noted. Exterior finishes (galvanizing, paint) should be inspected for evidence of degradation. The structural condition of anchor bolts or embedded plates used to secure the platform members should also be inspected and noted.
- Access doors – All components of mandoors and roll-up doors should be inspected for their condition and operation. Doorframes and anchorage to the concrete shell or concrete liner should also be inspected. Door hinges, latches, and seals (particular in pressurization systems) should be inspected and their condition noted.
- Tension bands – Each exterior band and band splice connection should be inspected for severity of corrosion and condition of the splice connection (bolted or welded). The condition of the weatherproofing caulk along the top of each band should also be noted.
- Breeching duct - All accessible surfaces of the duct plate, stiffeners, insulation, and perimeter of the duct penetration into the concrete shell or liner should be inspected and condition noted. Any turning vanes located within the breeching duct should also be examined for their condition.
- Lightning Protection System – Specifics on the quantity, location, and condition of items such as air terminals, downleads, lead covering on the uppermost 25' of system, anchors, and splice connections should be noted.
- Aviation Warning Obstruction Lighting (AWOL) - Specifics on the quantity, location, and condition of each aviation light, light bracket or door, and associated wiring or electrical components should be noted.
- Annulus pressurization system – The general condition of the pressurization fan, fan motors, and ductwork into the concrete shell should be inspected. All opening seals in the concrete shell should be inspected for holes, tears

and gaps. The anchorage of all seals to the concrete shell and support bars should be inspected.
- Elevator – The general condition of all accessible elevator components should be inspected and noted. Qualified elevator technicians should perform thorough inspections of chimney elevators at regularly scheduled intervals.

3.4.3 Class III Inspection

A Class III inspection, as identified in the introduction, is not routine and in the case of a concrete shell or liner, will typically be required only when significant degradation or an unusual event has occurred, or when a change in design such as installing a new flue opening or extending the height of the structure is being considered. The height and diameter of the concrete shell or concrete liner will dictate the number of external and internal inspection drops required to evaluate fully the condition of the structure.

In such a case, a Class III inspection should include all items outlined in the Class II inspection as well as any additional monitoring, nondestructive testing or physical destructive testing deemed necessary by a qualified Engineer. In specific cases, this may include chemical analysis or physical testing to establish material composition and/or physical properties of the concrete and reinforcing. Examples of this type of monitoring or testing could include, but may not be limited to:

- Carbonation depth testing.
- Concrete condition and integrity assessment by Impulse Response test.
- Crack measurement and monitoring.
- Core sampling for petrographic examination or compression testing.
- Concrete strength testing via Swiss (Schmidt) Hammer or Windsor Probe.
- Concrete density testing via Ultrasonic Pulse Velocity (UPV) measurements.
- Potential corrosion activity measured using the half-cell potential method.
- Wall thickness measurements using the impact echo test.
- Ultrasonic thickness measurements of duct plate, annulus rainhood or cap, turning vanes.
- Hazardous material testing of coatings.

CONCRETE SHELL OR LINER INSPECTION CHECKLIST

On plan sketch at right, indicate:

- North Arrow
- Flue Opening
- LPS Downlead(s)
- Ladder(s) and/or Manlift
- Construction Opening
- Access Opening

- Aviation Warning Light System (show elevation(s), quantity, size, orientation, color and type)
- Breeching duct(s) location (show size, orientation and sill elevation)
- Exterior tension bands (show size, quantity and elevation)
- Lightning protection system (show quantity and location of air terminals and downleads)
- Ladder(s) (show range of elevation and orientation, and specify if caged)
- Manlift (Show orientation and range of elevation)
- Platforms (show elevation and width)
- Cellular antennae (show quantity, size and orientation)
- Construction and access (mandoor) openings (show elevation, orientation and size)
- Equipment monorail beams (show elevation, orientation and projection)
- Sampling ports (show elevation, orientation and size)
- Any other areas of concern that require immediate or future maintenance

NORTH   EAST   SOUTH   WEST

Top Quarter

Third Quarter

Second Quarter

Lower Quarter

On the Chimney outlines above, sketch any or all of the following features – clarify if exterior or interior:

- Cracks (Show location, orientation and width)
- Significantly spalled or honeycombed areas of concrete

Additional Comments _____
_____
_____
_____

Safety Hazards (Existing and/or Potential) _____
_____
_____
_____

CONCRETE SHELL OR LINER INSPECTION CHECKLIST

Inspection Date: _____

Company Performing Inspection: _____

Owner Name: _____

Owner Address: _____

Plant Name: _____

Plant Contact: _____

Plant Address: _____

Send Report to attention of: _____

Address: _____

Construction Type (Jumpform, slipform): _____

Chimney Height: _____ Form Height: _____

Top OD: _____ Top Wall Thickness: _____

Base OD: _____ Base Wall Thickness: _____

Chimney Cap or Rainhood Material _____

Condition _____

Condition of Concrete Surface Throughout Exterior: _____

_____

_____

Condition of Concrete Throughout Interior: _____

AWL System?: ☐ No ☐ Yes : Elevations: _____

Number of Lights per Elevation: _____

Type of Lights: Steady Red ☐ Flashing Red ☐

High Intensity White ☐ Medium Intensity White ☐

All AWLS lights Operational? Yes ☐ No ☐ : List

AWL System: Condition of Conduits/Materials/Anchorage

_____

Exterior Coating: No ☐ Yes ☐ - Type: _____

Elevation Range _____

Coating Condition: _____

_____

Breeching Duct(s) - Qty: _____ Orientation: _____

Sill Elevation(s): _____

Breeching Duct(s) Size (Width x Height) _____

Breeching Duct(s) Condition _____

Exterior Lagging/Insulation _____

CONCRETE SHELL OR LINER INSPECTION CHECKLIST

Ladder(s)? No ☐ Yes ☐ : Interior ☐ Exterior ☐

Caged? No ☐ Yes ☐ Elevation Range _____

Ladder Condition: _____

Saf-T-Devices : No ☐ Yes ☐

Manlift?: No ☐ Yes ☐ : Interior ☐ Exterior ☐

Elevation Range: _____ No. of Stops: _____

Exterior Platform(s)? No ☐ Yes ☐

Elevation(s): _____

Width: _____ 90° ☐ 180° ☐ 360° ☐

LPS: No ☐ Yes ☐ - Lead Cover at Top? No ☐ Yes ☐

Qty Points: _____ Qty Downleads _____

LPS Condition: _____

_____

_____

Exterior Band(s) No ☐ Yes ☐ - Qty: _____

Elevations: _____

Band Condition: _____

_____

Construction Opening Size _____

Existing door? No ☐ Yes ☐ - Condition: _____

Access (Mandoor) Size: _____

Existing door? No ☐ Yes ☐ - Condition: _____

_____

Annulus Pressurization? No ☐ Yes ☐

Condition of Fans/Ducts/Seals: _____

_____

Chimney Originally Built by _____

Year Constructed: _____ Existing Drwgs: No ☐ Yes ☐

When was last repair? _____

By Whom? _____

Scope of work performed: _____

_____

_____

_____

_____

_____

_____

4.0 BRICK SHELLS

4.1 General Information

4.1.1 Structure Description

This section specifically addresses the inspection of brick shells. The brick shell should be considered any structure constructed of hollow or solid masonry units, manufactured from clay or shale. The most common design under this category is a "traditional" radial brick chimney designed for industrial service, which is constructed of hollow (perforated) masonry units manufactured from shale or clay, which are formed to the radial lines of the chimney. Therefore, this section will utilize this type of structure as a prototypical example. Other designs, which are considered variations of the radial brick chimney, include:

- Common brick chimneys (circular and other shapes)
- Architectural or commercial building chimneys
- Tile or hollow block chimneys
- Reinforced masonry chimneys

4.1.2 Historical Background and Design Notes

A vast majority of the traditional radial brick chimneys were constructed prior to the 1960's, with many of the structures dating back to the early part of this century. As such, these structures were not designed and/or constructed to specific building codes or national standards; however, they were built to industry-accepted criteria, which historically had a proven track record. In general, radial brick shells are designed as dead weight structures using established cantilever beam formulas. The typical construction involves utilizing bricks of different depths in various combinations to construct wall sections of varying thickness. A bonded wall section is achieved by alternating the bond pattern and staggering vertical mortar joints to avoid "stacking" of mortar joints. The exterior surfaces are typically constructed plumb or with a uniform taper and wall section thickness changes are offset to the interior of the structure.

4.2 Construction Materials

4.2.1 Brick

In general, bricks used for the outer shell are manufactured of shale or refractory clay and can be classified as radial chimney brick, building brick or hollow tile.

The following is a description of each of these bricks:

a) Radial Chimney Brick

Radial chimney brick are classified as hollow unit masonry manufactured of shale or refractory clay and fired at temperatures of 2000°F. The outside faces of the brick are molded to a radius with standardized dimensions; however, several different sizes of brick are manufactured with varying depth. Perforations typically do not exceed 30% of the total cross sectional area of the brick and a crushing strength for a unit is normally within the range 4000 to 6000 PSI.

There is no specific ASTM standard for radial chimney brick, however, ASTM C 67; Specification for Sampling and Testing Brick and Structural Clay Tile is normally used in evaluating radial chimney brick.

b) Common Brick

Common (or building) brick are considered normal, right angled brick manufactured in standardized sizes, which are utilized in commercial building construction. These brick are typically required to comply with ASTM C 652, Standard Specifications for Hollow Brick (hollow masonry units made from clay or shale) or ASTM C 62, Standard Specification for Building Brick (solid masonry units made from clay or shale).

c) Tile or Hollow Block

Tile or hollow block are typically hollow units manufactured of structural clay or concrete masonry and although ASTM standards currently exist (i.e. ASTM C 126 and ASTM C 90), at the time the majority of these chimneys were constructed, no such standard was in place.

4.2.2   Mortar

In general, most brick shells are constructed utilizing a hydraulic setting Portland cement and sand mortar, as outlined in ASTM C 150, "Specifications for Portland Cement" or ASTM C 91, "Standard Specifications for Masonry Cement".

Mortar used for the original construction will typically include hydrated lime (which improves flexural characteristics), or in certain applications may be combined with small amounts of fire clay (to enhance thermal resistance).

4.3 Degradation/Modes of Failure

Inasmuch as the majority of industrial radial brick chimneys were constructed prior to current air quality standards and advancements in power generation technology (as well as at a time when fuel costs were considerably less), flue gas temperatures seldom dropped below the acid dew points. Consequently, it was not unusual for radial brick chimneys to be constructed without acid resistant linings or with only partial height linings at the base (often installed for thermal protection of the shell opposite the flue opening).

Further, considering that mortars typically used for this type of construction have poor acid resistance characteristics, these chimneys are particularly susceptible to corrosion related problems associated with acid condensation/attack when operating conditions change (i.e. boiler retro-fit, change in fuel type, etc.)

NOTE In most routine boiler operations, hydrochloric and sulfuric acids are the primary chemical compounds responsible for degradation. In the same way that it attacks concrete, the sulfuric acid will attack various compounds present in the Portland cement in the mortar (i.e. tricalcium aluminate and calcium hydroxide) and the resultant product forms gypsum. The formation of this by-product is also accompanied by a significant increase in volume (approximately 20%) and a considerable amount of pressure, which can produce a visible enlargement of the mortar joints. A swelling or bulging can often be observed in the structure and, since the strength of masonry in tension is relatively low, it is not unusual to see formation of stress cracks throughout the structure.

Weathering and freeze thaw cycles are also often responsible for degradation in colder climates.

4.4 Inspection Programs

4.4.1 <u>Class I Inspection</u>

a) External Visual Inspection

The exterior of the brick chimney should be visually inspected from existing access points (i.e. ladders, service platforms, elevators, etc.) when available. This external visual inspection should serve to examine the condition of the following areas or items closely:

— Structural cracks in the chimney wall or sections of the brick wall displaced out-of-plane.

- Mortar joints – Joints should be visually checked for cracked or open joints, dislodged repointing work, and degradation. Mortar joints should be probed to identify any softening and/or powdery chalk like appearance.
- Bricks – Bricks should be visually checked for surface degradation, cracks, crazing, spalls, and downwash or wind abrasion.
- Exterior tension bands – The quantity, location, and condition of exterior bands should be noted as well as that of the bolted or welded splice connection.
- Lightning protection system – Items such as air terminals, downleads, lead covering on uppermost 25' of system, anchors, and splice connections should be inspected for discontinuities, fraying, corrosion, and anchorage.
- Chimney cap or rainhood – If accessible, the condition of the steel or concrete cap should be inspected for degradation, broken components, and cracks.
- Breeching duct exterior plate (or lagging) and perimeter of penetration into brick shell – The duct should be inspected for missing or unsecured lagging, exposed insulation, and exterior degradation of the duct plate and stiffeners. The gap between the duct (plate or lagging) and brick wall should be inspected for an adequate seal to prevent flue gas leakage and rainwater entrance.
- Extended brick faces surrounding breeching duct penetration – The condition of bricks and mortar joints should be inspected and noted.
- Water tables at extended faces or transition walls – Cracked or dislodged sections of cement should be inspected and noted.
- Platforms and ladders – All members, bolted and welded connections, and finish on all materials should be inspected for overall structural condition. The condition of anchors used to secure members and brackets should also be inspected.
- Antennae (cellular) – The support mechanisms for all exterior antennae should be inspected for condition and anchorage to the brick shell.
- Gas sampling ports – The condition of the ports, flanges, cover plates, bolts should be noted. The perimeter of each penetration into brick shell should be checked for leakage and local cracking.
- Cleanout door – The condition of the door plate, frame, and hinges should be inspected.
- Interior floor at base of chimney – If possible, this should be performed by opening the cleanout door, the amount of buildup on stack floor should be estimated, and the lower elevations of the interior wall inspected.
- Exposed concrete foundation or pedestal – Cracks, spalled concrete or exposed rebar should be identified.

CHIMNEY AND STACK INSPECTION GUIDELINES 49

4.4.2 Class II Inspection

a) External Visual Inspection

The external visual inspection of the brick shell should include the examination of items identified in Section 4.2.1, performed from any existing access points, (i.e. ladders, elevators, service platforms) as well as from a full height transit (drop) approximately 180° from the ladder. The two full-height areas of the chimney (ladder side and 180° from the ladder) should be closely examined to determine the condition of the following areas or items:

- Cracks – Structural cracks should be identified and mapped onto a chimney development sketch. The width, height, elevation, and orientation of the full crack should be noted. Areas of the wall that are displaced out-of-plane should also be identified and described on the development sketch.
- Mortar joints – The depth of weathered or deteriorated (from abrasion) mortar, measured from the exterior surface of the brick wall, should be measured at elevations throughout the height of the chimney. The amount (percentage of surface area) of required joint repointing should also be quantified.
- Bricks – The quantity and location of spalled, deteriorated, crazed, or cracked bricks should be identified.
- Chimney cap or rainhood – The structural condition of the concrete or steel cap should be identified. Concrete caps should be examined for cracks, spalled concrete, degradation from abrasion, and exposed rebar. Steel (cast iron, carbon or stainless steel) caps should be inspected for the condition of cap connections (bolts, welds) and anchorage to the top of the chimney.
- Tension bands – Each exterior band and band splice connection should be inspected for severity of corrosion and anchorage to the brick shell. The condition of the weatherproofing caulk along the top of each band should also be noted.

If possible, the orientation of a vertical drop should coincide with the location of the chimney breeching duct.

b) Internal Visual Inspection

The full height internal drop should provide direct physical access to the inner surface of the brick shell, internal liner, target wall, and/or baffle wall. The interior area of the chimney local to the vertical drop should be closely examined to determine the condition of the following areas or items:

- Structural cracks in the wall or sections of the wall displaced out-of-plane should be identified and mapped onto a chimney development sketch. The width, height, elevation, and orientation of the full crack should be noted.
- Mortar joints – Joints should be visually checked for cracked or open joints, dislodged repointing work, and degradation. Mortar joints should be probed to identify any softening and/or powdery chalk-like appearance. The depth of weathered or deteriorated (from abrasion) mortar, measured from the interior surface of the brick wall, should be measured at elevations throughout the height. The amount (percentage of surface area) of joint repointing necessary should also be quantified.
- Bricks – The quantity and location of spalled, deteriorated, crazed, cracked or damaged from downwash or wind abrasion.
- Baffle walls – The structural condition of the bricks, mortar joints, and connection to the interior surface of the chimney should be examined and noted on an interior development sketch.
- Buildup on interior surface – The amount (thickness) of buildup from ash, residue or combustion products should be measured at elevations throughout the height of the chimney.
- Corbel areas or interior shelves at wall thickness changes should be inspected for missing or cracked bricks. The level of buildup at these horizontal surfaces should also be noted. The condition of mortar washes at these locations should also be inspected and noted.
- Breeching duct – The interior plate, stiffeners, and perimeter of the duct penetration into chimney should be inspected. Any turning vanes located within the breeching duct or lintels located above the duct should also be examined for their condition.
- Flow straightening plates (egg crate design) – The structural condition of the plates, stiffening members, bolted or welded connections, and anchorage to the interior wall should be examined.
- Gas sampling ports – The condition of the ports, flanges, cover plates, bolts should be noted. The perimeter of each penetration into brick shell should be checked for leakage and local cracking.
- Interior openings – The condition of the bricks and mortar joints local to the breeching duct and cleanout door openings should be examined. Lintel beams should be checked for degradation, rust and bearing support.
- Interior floor – The amount (thickness) and hardness of buildup from ash, residue or combustion products should be measured and noted.
- Coatings – Exterior coatings should be inspected in accordance with Section 9.0 of this guide.

If possible, the orientation of the interior drop should coincide with the location of the chimney breeching duct.

4.4.3 Class III Inspection

A Class III inspection, as identified in the introduction, is not routine and in the case of a brick chimney shell, will typically be required only when significant degradation or an unusual event has occurred or when a change in design such as installing a new flue opening or extending the height of the structure is being considered.

In such a case, a Class III Inspection should include all items outlined in the Class II Inspection as well as any additional monitoring or physical destructive testing deemed necessary by a qualified Engineer. In specific cases, this may include chemical analysis or physical testing to establish material composition and/or physical properties of the brick and mortar. Examples of this type of monitoring or testing could include, but may not be limited to:

− Structural crack monitoring.
− Survey analysis to determine amount of vertical lean.
− Petrographic examination.
− Ultrasonic thickness measurements of duct plate, turning vanes, or flow straightening devices.

BRICK CHIMNEY INSPECTION CHECKLIST

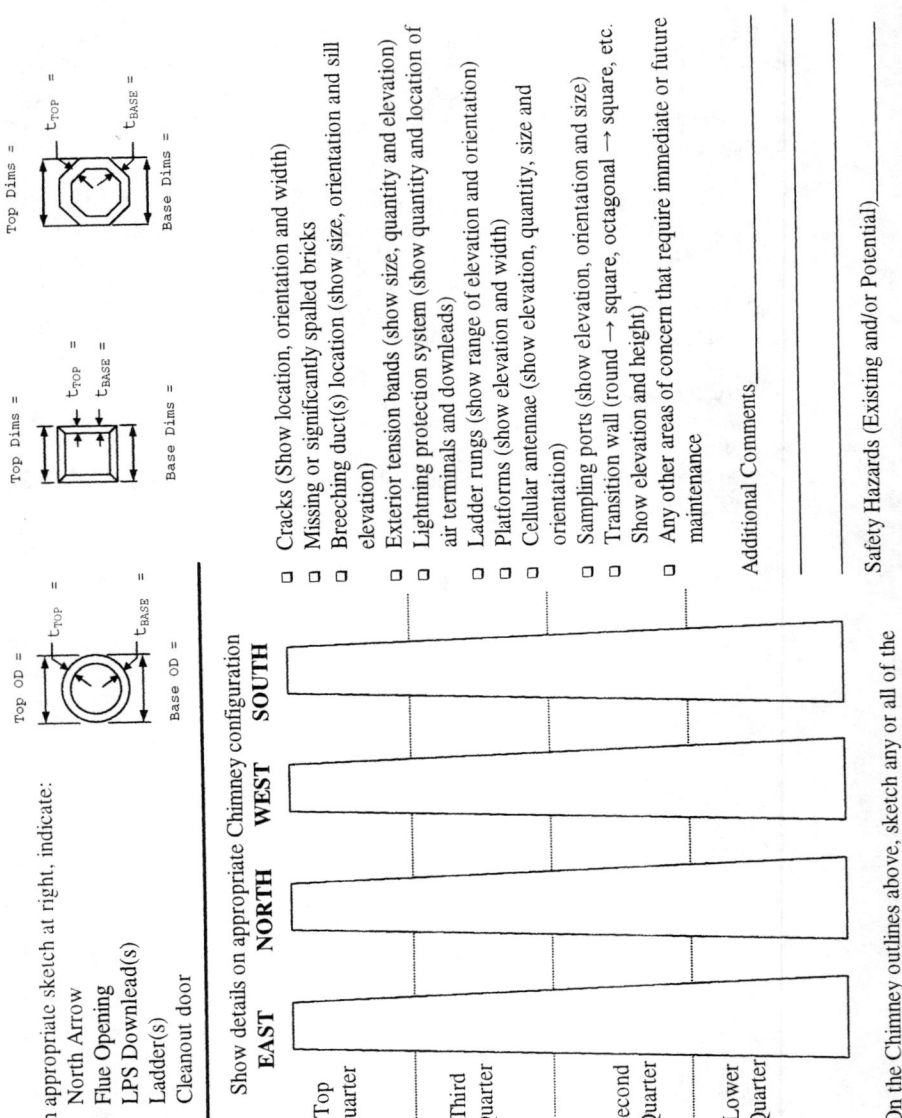

On appropriate sketch at right, indicate:
- ☐ North Arrow
- ☐ Flue Opening
- ☐ LPS Downlead(s)
- ☐ Ladder(s)
- ☐ Cleanout door

Show details on appropriate Chimney configuration
EAST   NORTH   WEST   SOUTH

Top Quarter

Third Quarter

Second Quarter

Lower Quarter

- ☐ Cracks (Show location, orientation and width)
- ☐ Missing or significantly spalled bricks
- ☐ Breeching duct(s) location (show size, orientation and sill elevation)
- ☐ Exterior tension bands (show size, quantity and elevation)
- ☐ Lightning protection system (show quantity and location of air terminals and downleads)
- ☐ Ladder rungs (show range of elevation and orientation)
- ☐ Platforms (show elevation and width)
- ☐ Cellular antennae (show elevation, quantity, size and orientation)
- ☐ Sampling ports (show elevation, orientation and size)
- ☐ Transition wall (round → square, octagonal → square, etc. Show elevation and height)
- ☐ Any other areas of concern that require immediate or future maintenance

Additional Comments _____

On the Chimney outlines above, sketch any or all of the following observed features:

Safety Hazards (Existing and/or Potential) _____

BRICK CHIMNEY INSPECTION CHECKLIST

Inspection Date _____ Condition of Mortar Joints Throughout Exterior _____
Company Performing Inspection _____ _____
Owner Name _____ _____
Owner Address _____ Condition of Mortar Joints Throughout Interior _____
_____ _____

Plant Name _____ _____
Plant Contact _____ Condition of Bricks Throughout Exterior _____
Plant Address _____ _____
_____ _____
_____ Condition of Bricks Throughout Interior _____

Send Report to _____ _____
Address _____ _____
_____

Chimney Height _____ Top OD _____ Support: ☐ Free Standing ☐ Roof Supported:
Top Wall Thickness _____ Height Above Roof _____ Height Below Roof _____
Type of Brick (Radial, Buff, Common) _____ Interior Lining ☐ No ☐ Yes : Type _____
Color _____ Chimney Cap Material _____ Elevation Range _____
Condition of Cap _____ Lining Condition _____
_____ _____

BRICK CHIMNEY INSPECTION CHECKLIST

Target Wall : ☐ No ☐ Yes : Type _____   Exterior Platform(s) ☐ No ☐ Yes : _____

Elevation Range _____   Elevation(s) _____ Width _____

Breeching Duct(s): Qty ___ Size (Width x Height) ___   ☐ 90° ☐ 180° ☐ 360° _____

Sill Elevation _____   LPS ☐ No ☐ Yes : Qty Points ___ Qty Downleads ___

Breeching Duct Condition _____   Lead Cover at Top : ☐ No ☐ Yes

_____   LPS Condition _____

_____   _____

Breeching Duct Opening Extended Face Condition _____   Exterior Band(s) ☐ No ☐ Yes : Qty _____

_____   Elevations _____

_____   _____

_____   Band Condition _____

Watertable(s) ☐ No ☐ Yes : Elevation(s) _____   _____

Condition _____   _____

_____   _____

_____   Cleanout Door : Size _____ Sill Elevation _____

_____   Soot Depth at Base _____

Outside Ladder ☐ No ☐ Yes : Type _____   Chimney Originally Built by _____ Year _____

Elevation Range _____ Saf-T-Devices: ☐ No ☐ Yes   Existing Drwgs: ☐ No ☐ Yes

Ladder Condition _____   When was last repair? _____

_____   By Whom ? _____

_____   _____

5.0 STEEL AND ALLOY STACKS

5.1 General Information

The periodic inspection of steel stack shells, flues, coatings, linings, appurtenances and other related stack system components is necessary for safety and to prolong the stack service life.

A complete inspection (Class II) can only be accomplished with the stack system out of service (off-line); however, a periodic on-line inspection (Class I) can be a helpful guide to when an off-line inspection is warranted. Depending on the service conditions, a steel stack may need to be on-line inspected more frequently and off-line inspected annually with a dated record kept on file.

After a potentially damaging event a class III inspection should be performed immediately with the stack still on-line, if necessary, followed by a class II off-line inspection as soon as possible.

5.2 General Descriptions

5.2.1 Steel Stack Types

Steel Stacks are designated by type:
- Single wall
- Dual wall with internal flue
- Multi-flue
- Cluster

5.2.2 Steel Stack Support

Steel stacks may be:
- Self-supporting at grade or on a structure
- Braced to a building or a structure
- Guyed with cables

5.2.3 Corrosion Protection

Steel stacks are usually protected against corrosion externally, and sometimes internally, by:
- Painting
- Galvanizing
- Elastomeric coatings
- Refractories
- Alloy cladding
- Special coating systems

5.2.4 Heat Transfer Control

Steel stacks transfer heat readily and are often provided with some means to control heat loss and thereby maintain flue gas temperatures throughout. Control may be by:
- External insulation and lagging on single wall stacks
- Insulation of flue(s) or inner wall of shell on dual wall and multi-flue Chimneys
- Internal insulating refractory
- Internal insulation protected by coating or metal lining

5.2.5 Appurtenances

Steel stacks are often fitted with appurtenances for access and inspection, or for other specific purposes. These may include:
- Ladders with Saf-T Climb devices or cages
- Platforms for testing and instrumentation
- Platforms for access, maintenance, or rest
- Aviation Warning Obstruction Lighting
- Platform and ladder lighting for safety
- Access and maintenance doors
- Test and instrument ports
- Lightning grounding lugs and cables
- False bottom with drain
- Exit cone
- Silencers
- Internal dampers
- Roof flashing and counter-flashing
- Top rainhoods or shrouds
- Turning or straightening vanes
- Platform jib cranes or hoists
- Dynamic stability devices including mass dampers, fins, shrouds or strakes, and base to foundation resilient pads

5.3 Construction Materials

5.3.1 Steel Stack Shells

The shell (structural component) of a steel stack may be fabricated of:
- Carbon steel
- Self-weathering carbon steel (not recommended for wet service)
- Stainless steel
- Nickel alloy steel
- Clad plate steel

5.3.2 Steel Stack Flues

The flues (the components designed to carry process gases or exhaust air) of steel stacks may be designed and fabricated of:
- Carbon Steel
- Self-weathering carbon steel (not recommended for wet service)
- Stainless steel
- Nickel alloy steel
- Titanium
- Clad plate steel
- Glass fiber reinforced plastic (FRP)

5.3.3 Insulation

The external or internal insulation (dual wall or multi-flue) for steel stacks is usually designed and installed over pin studs, and sometimes reinforced with wire mesh, using:
- Fiberglass Thermal Insulating Wool (TIW)
- Mineral wool

5.3.4 Lagging

The protective material applied over the external insulation on steel stacks may be designed to be fabricated of:
- Aluminum
- Stainless steel
- Galvanized carbon steel
- Painted carbon steel
- Plastic

The lagging is usually held in place by fasteners and banding to accommodate differential expansion.

5.3.5 Coatings and Linings

See Section 9.

5.4 Degradation/Modes of Failure

5.4.1 Corrosion

By far the most significant failure potential for steel stacks and their flues is through corrosion. Protective coatings are an absolute necessity to ensure a long service life. See section 9.

5.4.2 Stress or Fatigue Cracks

Steel stacks and their flues are subject to stress and fatigue cracks due to repetitive or excessive movement caused by either down-wind or across-wind forces. Cracks usually develop at openings or discontinuities in the metal shell.

5.4.3 Buckling

Steel stacks may fail through buckling of the wall cylinder after thinning of the shell material caused by corrosion.

5.4.4 Overheating

Steel stacks may deteriorate due to overheating of the shell or flue material. This degradation may be localized, as a result of partial loss of thermal insulation or protective coating.

5.4.5 Differential Expansion

Differential movement between components that are subjected to different temperatures, and are not free to expand or contract independently may cause steel stack damage.

5.4.6 Unusual Occurrence

Steel stacks may be damaged by an unusual occurrence such as fire, explosion, implosion, projectile impact, or support structure failure.

5.5 Inspection Programs

An inspection checklist and record form is provided in the Appendix to guide a steel stack owner in the stack system components to be covered in a thorough inspection. A checklist can be customized for each individual stack. It is best if the same inspector is used for each periodic Class I inspection to assure continuity and consistency of documentation. Inspection procedures described in this manual should serve as guide to the inspector.

Evidence of corrosion is the main indicator of degradation. Other key indicators to look for are cracks and broken welds. Specific items to document in the inspection should include the following:

5.5.1 Class I Inspection

a) Bolts and Nuts
— Anchor Bolts - Degradation is caused by corrosion located between the stack base plate and the concrete foundation. The covering of grout makes

visual inspection impossible. If cracks are observed in the grout around the anchor bolts or moisture is observed at the base plate, removal of the grout around one anchor bolt to inspect and measure its diameter is recommended. A 25% reduction from the original diameter should be considered cause for immediate remedial action.
- Anchor Bolt Nuts – inspect for corrosion.
- Horizontal Splice Bolts And Nuts - A bolt and nut should be removed for closer inspection if degradation is suspected.
- Platform and Ladder Anchor Bolts and Nuts - A bolt and nut should be removed for closer inspection if degradation is suspected.
- Any Other Bolts and Nuts - Including fixing of appurtenances. A bolt and nut should be removed for closer inspection if degradation is suspected.

b) <u>Shell and Components</u>
Inspect the following for evidence of corrosion, and the additional items listed:
- Base Plate
- Anchor Chairs
- Breeching Opening Reinforcing - also look for cracks or broken welds
- Shell Plate - Loss of plate thickness is best found by ultrasonic testing. Also look for any local buckling, and inspect vertical and horizontal welds for cracks
- Circumferential Stiffeners - also look for any cracks or broken welds
- Top Cone
- Roof - also look for broken welds
- Lateral Supports - also look for any bending or broken welds
- Access/Inspection Doors - A general condition inspection is required.
- False BottomDrain - also look for blockage or plugging
- Dynamic Stability Device - A general condition inspection is required. Refer to design drawings.
- Test and Instrumentation ports – look for evidence of leakage, corrosion and/or cracking.

c) <u>Appurtenances</u>
- See Section 10

d) <u>Insulation and Lagging</u>
- A general condition inspection is required on external insulation and internal flues, if accessible. Moisture penetrating the lagging is the main concern.

e) <u>Expansion Joints and Seals</u>
- A general condition inspection is required of both the bellows (or fabric) and the frames. Corrosion and cracks or holes in the fabric are the main concerns.

f) Grounding Cables
— A general condition inspection is required. Look for evidence of corrosion, fraying or chafing.

g) Guy Wires
— Cable degradation and broken wire strands are the main concern.

h) Guy Wire Dead Men
— Cable clamps and anchors require inspection for breaks or corrosion.

5.5.2 Class II Inspection

a) Flue Plate
Corrosion is the main concern. Loss of plate thickness is best found by ultrasonic testing. Look for any local buckling. Inspect vertical and horizontal welds for cracks.

b) Flue Inlet Duct
Corrosion is the main concern. Loss of plate thickness is best found by ultrasonic testing. Look for any local buckling. Inspect vertical and horizontal welds for cracks.

c) False Bottom
Corrosion is the main concern.

d) Drain
Plugging and corrosion are the main concerns.

e) Flue Supports
A general condition inspection is required.

f) Lateral Supports
Plugging and corrosion are the main concerns.

g) Exit Cone
Corrosion is the main concern.

h) Shell External Coating
A general condition inspection is required. Look for flaking.

CHIMNEY AND STACK INSPECTION GUIDELINES 61

i) Shell Internal Lining
A general condition inspection is required. Refer to design drawings and coating or lining manufacturer's recommendations. Look for cracking or spalling of refractory type lining.

j) Flue External Coating
Generally, the external surface of the flue is not accessible on steel stacks due to insulation. It may be necessary to remove a section of insulation. A general condition inspection is required.

k) Flue Internal Lining
A general condition inspection is required. Refer to design drawings and coating or lining manufacturer's recommendations.

5.5.3 Class III Inspection

In addition to all items from Class I and II that can be performed with the stack on-line, identify and analyze any specific areas of visible damage to the stack or components which affect structural safety or warrant attention to corrosion abatement. Describe the severity of the damage, and assess the urgency of necessary repairs.

CLASS I INSPECTION OF STEEL STACKS AND FLUES

SAMPLE CHECKLIST AND RECORD FORM

Stack Designation:_____

Date of Inspection: _____ Ambient Conditions:_____

Inspector's Name: _____

APPURTENANCES **GENERAL CONDITION**

1. Ladder/Stairs:	
2. Platforms:	
3. Test Ports	
4. Instrument Ports	
5. Aircraft Lighting	
6. Electrical and Lighting	

Special Comments:_____

_____

_____

_____

OTHER **GENERAL CONDITION**

1. Insulation	
2. Insulation Lagging	
3. Expansion Joints & Seals	
4. Grounding Cable	
5. Guy Wires/Dead Men	

Special Comments:_____

_____

_____

_____

CLASS II INSPECTION OF STEEL STACKS AND FLUES

CHECKLIST AND RECORD FORM

Stack Designation: _____

Date of Inspection: _____ Ambient Conditions: _____

Inspector's Name: _____

BOLTS AND NUTS   **GENERAL CONDITION**

1. Anchor Bolts and Nuts	
2. Flange Bolts and Nuts	
3. Platform and Ladder Bolts and Nuts	
4. Other Bolts and Nuts	

Special Comments:_____

SHELL   **GENERAL CONDITION**

1. Base Plate:	
2. Base Chairs:	
3. Breeching and Opening:	

4. Shell Plate:
     Bottom:
     Middle:
     Top:	

5. Circumferential Stiffeners:	
6. Top Cone:	
7. Roof:	
8. Lateral Supports:	
9. Access/Inspection Doors:	
10. False Bottom:	
11. Drain(s):	
12. Dynamic Stability Device:	

Special Comments:_____

CLASS II INSPECTION OF STEEL STACKS AND FLUES

CHECKLIST AND RECORD FORM

Stack Designation: _____

Date of Inspection: _____ Ambient Conditions: _____

Inspector's Name: _____

FLUE (S)   **GENERAL CONDITION**

1. Flue Plate	
2. Flue Inlet	
3. False Bottom	
4. Drain(s)	
5. Flue Support(s)	
6. Lateral Support(s)	
7. Top Cone	

Special Comments: _____

_____

_____

COATINGS AND LININGS   **GENERAL CONDITION**

1. Shell External:	
2. Shell Internal:	
3. Flue External:	
4. Flue Internal	

Special Comments: _____

_____

_____

_____

CLASS I, II AND III INSPECTION OF STEEL STACKS AND FLUES

CHECKLIST AND RECORD FORM

Stack Designation: _____

Date of Inspection: _____ Ambient Conditions: _____

Inspector's Name: _____

IMMEDIATE REPAIR RECOMMENDATIONS:

_____
_____
_____
_____
_____
_____
_____

Signed: _____

6.0 BRICK LINERS

6.1 General Information

6.1.1 Structure Description

This section specifically addresses the inspection of brick liners. Brick liners should be considered to be any structure within the chimney shell that is constructed of hollow or solid masonry units, manufactured from clay or shale. The three most common designs for brick liners are:

a) Free-standing column (circular and other shapes, constructed independent of outer shell)
b) Contiguous wall liner (in contact with and usually tied to outer shell)
c) Corbel-supported sectional liner.

While all brick liners share certain features in common, each of these designs has specific characteristics that require consideration when planning and performing an inspection. These are discussed in the following text.

6.1.2 Historical Background and Design Notes

The majority of brick lined chimneys were designed and constructed to operate as natural-draft convection chimneys, where the high temperature differential between the flue gas and the ambient air created strong convection currents that expelled the flue gases with sufficient velocity to ensure good atmospheric dissipation. In such a system, the gas pressure within the flue is slightly lower than ambient air pressure, so any cracks or openings in the liner will result in an in-flow of fresh air from the outside.

The installation of precipitator or scrubber systems in recent years has resulted in lower flue gas temperatures, and consequently lower gas ejection velocity at the top of the chimney, and poorer gas dissipation. Booster fans have been installed to achieve the necessary ejection velocity, with the result that the gas pressure in the liner system is slightly above ambient atmospheric pressure, and any cracks or openings in the liner system result in flue gas leakage through the liner. Where these gases impinge on the outer shell or other parts of the structure, the temperatures often fall below the dew-point of the acids in the gas stream, resulting in acid condensation and chemical attack.

In general, up until the late 1970's brick liners were designed as dead weight structures. Liners built since the 1970's are designed for both dead load and axial bending considerations, in accordance with ASTM C-1298. That

CHIMNEY AND STACK INSPECTION GUIDELINES 67

document contains a wealth of information on brick liner systems. As for brick shells, typical brick liner design involves utilizing bricks of different depths in various combinations to construct wall sections of varying thickness. A bonded wall section is achieved by alternating the bond pattern and staggering vertical mortar joints to avoid "stacking" of mortar joints. Unlike brick shells, liner interior surfaces are typically constructed plumb or with a uniform taper and wall section thickness changes are offset to the exterior of the structure.

6.2 Construction Materials

6.2.1 Brick

In general, bricks used for the construction of older liners are similar to those used in the construction of brick chimney shells - Refer to Section 4.0 for more details. Brick liners constructed later may be composed of acid-resistant brick or glazed brick, bonded with acid resistant mortar.

6.2.2 Mortar

Most older brick liners, like brick shells, were constructed using a Portland cement and sand mortar. (See Section 4.2.2) Mortar used for the original construction may have included hydrated lime to improve flexural behavior or, in certain applications, may be combined with small amounts of fire clay to enhance thermal resistance.

Newer brick liners are generally constructed with acid resistant mortar, which is characterized on site by much thinner mortar layers than with conventional cement mortars.

6.2.3 Reinforcing Bands

Contiguous brick liners are typically tied into the outer shell and are unreinforced. Independent and corbel-supported liners are typically reinforced with horizontal steel bands around the exterior of the liner. These bands are typically spaced vertically about every four feet.

6.3 Degradation/Modes of Failure

6.3.1 Degradation of Mortar

The mortars typically used for the older brick liners were intended to be exposed to flue gases at temperatures well above the dew-points of the acids in the gas stream. These mortars have poor acid resistance characteristics, making the liners highly susceptible to chemical attack and erosion resulting from acid condensation when changed operating conditions reduce flue gas temperatures (i.e. precipitators, scrubbers, changes in fuel type, etc.).

Acid attack on the mortar can result in swelling of the mortar joints (see Section 4.3). Swelling or bulging can often also be observed in the liner, sometimes accompanied by stress cracks that may penetrate right through the structure. Some forms of acid attack result in softening or crumbling of the mortar.

6.3.2 Degradation of Bricks

The lower temperature flue gases often have higher moisture content, which can be absorbed by the liner bricks, particularly those opposite the flue breeching. These bricks also swell as they absorb the moisture, and can create a significant "lean" in an independent liner.

Stripping or scaling of the brick faces can result from either acid attack or from excessively high flue gas temperatures.

6.3.3 Degradation of Flue Ducts and Reinforcing Bands

On independent or corbel-supported liners, distortion or sagging of the flue breeching duct and lintel can result from build-up of flyash on top of the flue duct in the annular space. Thermally induced movement of the duct lining relative to the brick liner can result in the loss of packing or sealing materials. Either situation allows flue gases and ash to enter the annular space.

Corrosion and breakage of reinforcing bands can result from flue gases leaking through cracks or breechings of the liner.

6.3.4 Blockage of Drains

Liners are typically provided with base drains or vents to prevent accumulation of water or flyash sludge from rain ingress or other sources, such as scrubber leaks or condensation. If these drains become plugged, water or sludge may accumulate to significant depths, imposing lateral loads on the liner that can cause failure. Brick liners often have elevated floors, and are therefore also susceptible to damage from vertical loads caused by plugged drains.

6.4 Inspection Programs

6.4.1 Class I Inspection

a) External Visual Inspection

On contiguous or corbel-supported liners, no external visual inspection is possible. On some independent liners, access may be possible to the annular space when the chimney is online. On these chimneys the annular space should be inspected at least every six months to check for build-up of fly-ash deposits, fallen masonry, or evidence of broken reinforcing bands.

Liner drains should be checked for plugging or accumulation of sludge, particularly in liners with elevated floors.

A Class 1 inspection should be performed on independent liners every 6 to 24 months, depending on recorded chimney condition. Observations should include review of as much of the liner exterior as possible, identification of the locations of any missing or broken reinforcing bands, and a visual check of the plumbness of the liner, and its concentricity within the shell at the top of the chimney. An internal inspection should be performed when accessible.

6.4.2 Class II Inspection

a) External Visual Inspection

External visual inspection of an independent brick liner should include examination of all items identified in Section 6.2.1, performed from any existing access points, (i.e. base, access ports, ladders, elevators, annulus platforms); as well as from at least one transit (drop) over as much of the height as the annular space permits. On larger diameter chimneys two transits (drops) approximately 180° from each other may be desirable.

The condition of all accessible reinforcing bands should be noted, and the location of any missing or broken bands documented.

b) Internal Visual Inspection

All Brick Liners

Typically, the interior surface of the liner should be cleaned of all adherences of flyash prior to performing the visual inspection. However, caution should be used when specifying cleaning methods, since frequent cleaning by high-pressure washing can be detrimental under certain circumstances. Advice on the frequency and advisability of high pressure washing should be sought from an experienced consultant who is familiar with the chimney in question.

The full height internal drop(s) should provide direct physical access to the inner surface of the brick liner, and at least one drop should be oriented to permit a close-up examination of the flue opening. Close attention should be paid to the seal between the flue duct material and the chimney liner.

The brick faces should be examined for evidence of scaling or flaking, and the mortar seam should be probed with a metal pick or scraper to check for softening or crumbling.

The location, width, and length of all cracks should be accurately noted for comparison with earlier or later inspection reports.

Corbel Supported Brick Liners

Particular attention should be paid to the horizontal liner joints at each corbel. These joints typically include an overlap section of specially formed brick. The overlap section forms a sliding joint that accommodates movement caused by thermal expansion and contraction of the shell and liner system.

The inner lip of the joint must be packed with a suitable flexible material (rope or elastomeric sealant) to prevent fly-ash ingress and build-up that inhibits movement of the joint. If movement of the joint is restricted, buckling and failure of the liner section or corbel ledge can occur.

The location of reinforcing bands can be confirmed nondestructively using a metal detector or subsurface interface radar (SIR).

6.4.3 Class III Inspection

A Class III inspection, as identified in the introduction, is not routine and in the case of a brick liner, will typically only be required when significant degradation or an unusual event has occurred; or when a change in design such as installing a new flue opening or extending the height of the structure is being considered.

In such a case, a Class III Inspection should include all items outlined in the Class II Inspection as well as any additional physical sampling and destructive testing deemed necessary by a qualified Engineer.

In specific cases, this may include chemical analysis, or physical testing to establish material composition and/or physical properties of the brick and mortar.

7.0 STEEL AND ALLOY LINERS

7.1 General Description

7.1.1 Structure Description

Section 7 specifically discusses the inspection of steel chimney liners; typical of those referred to in the American Society of Civil Engineers publication on Design and Construction of Steel Chimney Liners. In general, this refers to single flue and multi flue arrangements, which are characterized by a chimney that consists of an independent outer shell (normally constructed of reinforced concrete), an accessible airspace and an insulated steel liner(s).

The three most common steel chimney liner configurations are:

- Top Suspended Liner
- Base Supported Liner
- Floating/Toggle Liner

Although this section specifically deals with steel liners as described above, similar design concepts are applied on a smaller scale, with both steel and radial brick outer shells. The inspection of such structures should be performed in a similar manner to the procedures outlined in this section, to the extent possible considering that physical access to the air space and/or liners may be limited.

7.1.2 Liner Configurations

a) Top Suspended Configuration

The top suspended configuration normally involves suspending the entire liner from a location near the top of the reinforced concrete outer shell. Support grillage normally consists of structural members installed in beam pockets that have been formed in the reinforced concrete outer shell. This approach keeps the majority of the steel liner in tension, and all thermal expansion is downward. An expansion joint is typically included above the flues. Lateral restraints which permit radial and vertical expansion are typically included at specific locations throughout the height of the liner.

b) Base Supported Configuration

Base supported configurations typically involve either foundation supported designs, or incorporate a grillage support and ring girder installed above the flue openings. In either case, the steel liner is in compression and thermal expansion is vertically upward.

Lateral restraint in the form of stayrods, liner guides and/or bumpers is typically employed at specific locations throughout the height of the liner. Typically, however, minimizing lateral restraint is considered good industry practice.

c)   Floating/Toggle Configuration

The floating toggle arrangement, though less common, is an alternative design that typically incorporates the use of two expansion joints in the liner. The design is normally based upon installing the liner support grillage at a level above the flue gas mixing zone (typically 5 to 8 diameters), thereby reducing the effects of differential temperature stresses. The expansion joints are located immediately above the flue entry level and below the liner grillage support. Spring hangers typically support the floating toggle section of the liner.

7.1.3   Miscellaneous Design Components

a)   Grillage Support

Support designs are quite varied. However, systems normally include ring girder systems installed to the liner and structural members installed in beam pockets, supported at the reinforced concrete outer shell.

b)   Stayrods/Bumpers/Liner Guides

Various methods of providing lateral restraint to minimize deformation, reduce L/R and transfer horizontal forces while permitting thermal expansion have been employed in steel liners.

Some of the more common systems include tangential cables or tie rods, spring loaded pipe hangers, radial type bumpers and structural tube bumper guides.

c)   Insulation

Insulation provides both necessary and beneficial functions regarding the overall design of the chimney and steel liner. Maintaining steel liner surface temperatures above acid dew points is of critical importance in the operation of the steel liner, particularly in view of the fact that most steel liner designs include only a minimal corrosion loss factor.

In addition, the concrete column benefits in numerous ways from the reduced heat transfer through the liner.

7.2 Construction Materials

7.2.1 Liner

In general, most steel liners have been constructed of mild carbon steel, conforming to one of two specifications, ASTM A36 or ASTM A242 (often referred to as weathering steel). It has also been considered good industry practice to incorporate the use of a stainless steel within the upper portion of the liner (typically throughout the upper 25' to 50'). Stainless steels most frequently employed for this use include 316L and 304L, although higher nickel - chromium alloy stainless steels have been used in some instances.

The use of high alloy stainless steel or titanium alloys has grown in popularity as a "wall papering" material in retrofit applications and new steel liners for chimneys with flue gas desulfurization (FGD) systems.

7.2.2 Structural

Support structure, lateral restraints, stiffeners and other appurtenance related items are normally fabricated from carbon steel, conforming to ASTM A36.

7.2.3 Insulation

Insulation installed on the exterior surface of the steel liner and completely covering all structural attachments (i.e. stiffeners, girders, liner guides, etc.) normally consists of mineral wool type insulation installed over pins with speed clips. Industry practice recommends that insulation is a minimum of 2" thick, with a minimum density of 3.0 pcf.

7.2.4 Linings and/or Internal Coatings

Typically linings and/or internal coatings are not used unless flue gas temperatures are expected to fall below the acid dew point range.

7.3 Degradation/Modes of Failure

7.3.1 History

The use of steel liners became popular in the 1960's and 1970's, particularly in fossil-fuelled electricity generating stations. Despite this, until the formation of the ASCE Task Committee in February of 1972, no design code or recommended practice for the design and construction of steel liners existed. This committee was formulated after several bumper/stayrod and liner failures were identified in the United States.

The committee determined that most liner plate buckling was associated with bending stress induced by differential gas temperatures.

Further, the committee concluded that designs prior to 1972 did not sufficiently address the following:

- Gas temperature differences causing differential expansion and bending
- Positive and negative pressures
- Imperfections in construction
- Rapid temperature transients

These design-related deficiencies represent the most commonly observed modes of failure. However, after a number of years of operation and with the power generation industry emphasizing maximizing boiler efficiency, lower operating temperatures have increased concerns regarding corrosion of steel liners.

7.3.2 Thermal

As previously mentioned, thermal related failure modes are typically associated with bending stresses induced by differential temperatures and rapid temperature transients. These factors normally translate into failure modes most usually observed as liner plate buckling and/or cracking. In addition, liner stay rods or bumper failures have also been identified in specific instances in which the installation of too many stay rods prevented lateral movement.

7.3.3 Corrosion

Corrosion has not been a major mode of failure in steel liners, presumably because maintaining flue gas temperatures above acid dew points has always been considered an integral and well understood aspect of the design. However, this being said, it is becoming increasingly common to observe corrosion problems throughout the upper portion of the liner. Although these sections are normally constructed of stainless steel, fluctuating boiler load demands and the downstream proximity appear to create conditions in which flue gas temperatures drop below acid dew points.

In general terms corrosion is the electrochemical attack of metallic materials by reaction with their environment. A number of chemical compounds may be responsible for corrosion of steel liners in coal and fuel oil operations; however, the most common is associated with sulfurous acids. Burning fuel oil and coal typically produces sulfur or sulfuric gases in the form of SO_2 or SO_3. These compounds will oxidize under certain temperature and humidity conditions to form sulfuric acids (i.e. H_2S, H_2SO_3 and H_2SO_4). This process most commonly occurs when flue gas temperatures drop below the acid dew point, which, for most acids, is within the range 240° to 280° F. Within this service range corrosion may be a serious problem.

Also, corrosion related problems often occur when external insulation is installed improperly or damaged. This is frequently observed at attachments to the exterior surface (such as stiffener rings, appurtenances, stay rods, etc.), which tend to act as heat sinks when not installed properly.

7.4 Inspection Programs

7.4.1 General

The steel liner is designed as an integral unit in which the performance of the steel liner is greatly affected by the various components attached to it, i.e. support grillage, lateral restraints, insulation, etc. Furthermore, the performance of the steel liner will affect the overall performance of the outer reinforced concrete shell.

Consequently, it is of equal importance to perform a visual inspection of the steel liner(s), both internally and externally (from the air space), when accessible. The external visual inspection will be critical in terms of the examination of various liner components and the internal inspection will provide the opportunity to examine the condition of steel liner plates.

In addition to visual examination, the steel liner plate thickness should be monitored on a routine basis using an ultrasonic thickness gauge.

7.4.2 Class I Inspection

a) External Visual Inspection

The exterior of the steel liner should be visually inspected from existing access points (i.e. ladders, service platforms, elevators, etc.) when available. This external visual inspection should serve to examine the following closely:

− Concentricity and plumbness of the liner within the column
− Fly-ash deposits or other evidence of flue gas leakage through the liner.

- The steel liner including steel liner plates, insulation, stiffeners and other miscellaneous liner attachments.
- The support structures and related components, including grillage, ring girders and lateral restraints. Support grillage should be examined for deformation, over-stressing and/or general condition. The inspection should also include attachment/anchorage or support points at the reinforced concrete outer shell.
- The lateral restraint system(s) should also be examined for deformation, over stressing and/or general condition. Systems that provide for movement tolerances (i.e. tangential tie rods, spring loaded pipe hangers, etc.) should be examined carefully and the appropriate movement clearances should be noted and outlined in the report.
- All appurtenances, including items such as expansion joints, cap/rainhood, access doors, service platforms, ladders, sample ports, etc.

a)   Internal Visual Inspection

If the chimney is off-line, the interior of the steel liner should be visually inspected from one full height transit (drop). This internal visual inspection of the steel liner should serve to examine closely the condition of the steel liner plates, liner floor, expansion joints and breeching/liner connection.

In addition, specific attention should be given to the identification, orientation and size of any areas of distress, such as corrosion, deformed/buckled plate, cracked welds or other signs of over-stressing. The extent of surface corrosion and/or pitting should be evaluated and reported. Frequently, the presence of external attachments (i.e. stiffeners, support rings, lateral restraints, etc.) may result in heat-affected zones that are visually apparent on the interior of the steel liner. Any such areas should be identified and reported.

7.4.3   Class II Inspection

a)   External Visual Inspection

The external visual inspection of the steel liner should include the examination of items identified in Section 7.4.2-a performed from existing access points (i.e. ladders, elevators, service platforms), as well as from a transit (drop) 180° from the ladder and/or elevator, to the height permitted by the annular space.

The transit (drop) should provide direct physical access to the external surface of the steel liner and shall also be orientated to include a close-up examination of the lateral restraint mechanism as permitted.

b) <u>Internal Visual Inspection</u>

It is recommended that an internal visual inspection should include the examination of items identified in Section 7.4.2-b from four quadrants, located 90° apart, and minimum of two transits (drops) located at least 90° apart. More drops may be required for adequate coverage of larger liners.

c) <u>Ultrasonic Thickness Testing</u>

Steel plate liner thickness should be obtained using an ultrasonic thickness gauge. Ultrasonic thickness measurements should be obtained on four quadrants, located 90° apart and vertically on 10 ft intervals. Additional readings should be obtained adjacent to any areas of distress, as well as adjacent to areas noted as heat sink locations. (1)

7.4.4 <u>Class III Inspection</u>

A Class III inspection should include all items as outlined in a Class II inspection, as well as Non-Destructive Examination (dye penetrant method or x-ray examination) to check the integrity of critical welds. A critical weld should be defined as a highly stressed weld in shear or tension where failure may result in collapse of the liner.

Also, in specific cases, chemical analysis, mechanical testing and/or metallurgical examination of the steel plate may be warranted to establish material composition, physical properties and/or intergranular characteristics of the steel plate.

[1] Design and Construction of Steel Chimney Liners (ASCE Task Committee on Steel Chimney Liners, Fossil Power Committee Power Division - 1975.)

8.0 FIBERGLASS REINFORCED PLASTIC (FRP) LINERS AND STACKS

8.1 General Information

8.1.1 Structural Description

This section discusses the inspection of FRP chimney liners and stacks, typically those types referred to in the ASTM D 5364 "Standard Guide for Design, Fabrication, and Erection of Fiberglass Reinforced Plastic Chimney Liners with Coal-Fired Units." In general, this section refers to single or multi-flue arrangements which are supported by a reinforced concrete outer shell, skeletal structural steel frame, self supported, guyed or braced. These types of chimneys are usually equipped with an accessible airspace between the flue and the supporting structure. The use of FRP for the outer shell of a chimney/liner combination is rare.

For those considering changing existing lining systems to FRP, the Electric Power Research Institute (EPRI) has published EPRI Report No. TR-101654 - "Guidelines for the Use of Fiberglass-Reinforced Plastic in Utility Flue Gas Desulfurization Systems." Further information can be found online at www.epri.com.

FRP liners can be either top supported near the chimney's outlet and allowed to expand downward or bottom supported and allowed to expand upward toward the outlet. Usually FRP liners are not insulated.

FRP liners are assembled as cylindrical elements with circular cross-section, joined together with bolted, bell & spigot or butt joints. The FRP sections are usually filament wound and are typically shop-fabricated, but can also be fabricated in the field if necessary.

The FRP wall thickness is typically built-up in four basic layers. The inner, or base layer is resin rich and usually contains carbon black fillers. The second is also resin rich and contains a higher glass content than the first layer. The third layer is the structural wall and is a matrix of filament wound glass and chopped-strand. The fourth, or outer layer provides exterior protection from corrosive elements and ultraviolet light. In summary, the layers 1,2, and 4 provide corrosion protection for the structural layer.

8.1.2 Liner Configurations

Section 7.1.1 describes the top and bottom supported configurations for steel liners and need not be duplicated here. FRP liner configurations are similar.

In very tall chimneys, multiple supports and expansion joints may exist. Multiple supports are typically required to accommodate significant thermal expansion and strength limitations of the FRP shell.

8.1.3 Components

a) Breechings

An integral component of all liners and stacks is the flue ductwork or breeching. One end of the breeching is usually cantilevered directly off the liner and the other is fitted with an expansion joint.

b) Circumferential Stiffeners

Most FRP liners and chimneys are equipped with circumferential stiffeners which are integrally wound or bonded to the exterior surface of the liner. These may be fabricated of steel or FRP, and are used to improve hoop strength/stiffness.

c) Erection Lifting and Tailing Lugs

It is not uncommon for FRP liner sections to have erection lifting and tailing lugs which remain in place after construction. The lugs may be FRP or steel and some may span the joint between cylindrical sections. There are various designs.

d) Expansion Joints

Many liners have expansion joints positioned within the main cylindrical liner. The expansion joints vary in design and width. They are usually equipped with an interior slide, baffle or flow plate. Expansion joints may also have condensate drains and tie bars to limit movement and side sway.

e) Joints

FRP liner and stack sections are joined together using a variety of methods. The most common are flanged and bolted, bell and spigot, or square or tapered butt. The flanged and bolted connection will usually have a gasket or sealant at the mating surfaces.

The bell and spigot and the butt joints are sealed and structurally connected by wrapping the inside and outside faces with resin and glass. The wrap details vary depending on the strength requirements but in general are twelve inches wide and go all around the inside and outside surface of the joint.

f) Grounding

All liners and stacks should be equipped with a grounding system which prevents the dangerous build up of static electricity. The design concept is that ground lugs are integrally connected to the conductive carbon veil layer. The connection bolts of the lugs protrude to the exterior surface and copper ground cable connects the lugs to the chimney grounding and lightning protection system.

g) Pressurization Systems for Expansion Joints
Some expansion joints are pressurized and equipped with ductwork and fans. These systems are designed to pressurize and keep debris out of the expansion joint, thereby minimizing build-up and reducing the chance of binding.

h) Quench Systems
A quench system is included in the liner or stack design if there is a chance of a temperature excursion above the allowable liner design temperature. Water quenching is engaged at a preset shell temperature. Fine water droplets are injected into the exhaust gas stream via numerous nozzles around the circumference of the flue to reduce the gas temperature.

i) Rainhood
Chimneys with a single FRP liner are usually equipped with a FRP rainhood or weather- shield that covers the annular space between the exterior of the liner and the outer structure. The junction at the liner and rainhood is accomplished with an overlay of FRP or similar material.

8.1.4 Miscellaneous Design Components

a) Grillage Supports

Support designs are quite varied, but the majority of systems usually include a structural steel ring girder installed to the liner or stack and structural steel members underneath installed to bridge to the outer shell or skeletal structure.

b) Counter Weight Systems

Some liner designs incorporate an exterior counter weight system. The counter weight system reduces the dead load stresses in the liner and allows the designer to place more linear feet of the liner in compression.

c) Stayrods/Bumpers Guides

Various methods of providing lateral restrain to minimize deformation, reduce l/r and transfer horizontal forces while permitting thermal expansion have been employed. Some more common systems include tie rods, spring loaded pipe hangers, radial type bumpers and structural tube bumper girder. These are usually located near or connected to a FRP stiffener.

CHIMNEY AND STACK INSPECTION GUIDELINES 81

8.2 Construction Materials

8.2.1 Wall Construction (Typical)

The FRP liner or stack wall thickness is typically built up in four layers. The first layer is resin rich and lightly reinforced with "C" glass. Carbon fiber filler is usually added. The second is another thin resin rich corrosion resistant layer and has a higher glass content. The third layer is a thick structural layer of resin and glass filament which is wound around the liner. The fourth layer provides the exterior corrosion and UV protection.

8.2.2 Glass (Typical)

First Layer: Fine mat of type "C" glass plus a carbon filler
Second Layer: 1.5 oz. glass mat or a chopped strand mat
Third Layer: Type "E" Glass filament
Fourth Layer: None, fine mat or chopped strand

8.2.3 Resin (Typical)

Resin Types: Polyester or Epoxy Vinyl Ester
First Layer: 10 to 20 mils thick
Second Layer: 80 to 100 mils thick
Third Layer: As required by design but usually not less than 0.375"
Fourth Layer: 10 to 20 mils thick

8.3 Degradation/Modes of Failure

8.3.1 History

In the 1970's, the EPA published stricter clean air standards that required some utilities to scrub their flue gases. This led to cooler and more corrosive exhaust gases. Fiberglass Reinforced Plastic liners and stacks are capable of resisting the corrosive nature of these gases and also provide the necessary structural strength to withstand all other environmental loads.

FRP liners and stacks are capable of withstanding continuous operating temperatures up to 450 degrees Fahrenheit. The upper temperature limit depends on the resin selected. The chemical Antimony Trioxide is added to the resin to minimize the wall's flame spread ratio.

The first FRP liners were designed after the first generation of steel liners and the printing of the ASCE 1975 "Steel Liner Design" publication. Therefore, many of the problems initially found with steel liners did not occur in FRP liners. In addition, FRP liners are typically installed in tension. Therefore buckling has not been a problem since they generally experience lower gas temperatures, with consequent lower thermal stresses.

Repairs to FRP liners and stacks and their appurtenances have been needed because of :

- Surface weathering, erosion and corrosion
- Thermal
- Vibration
- Excessive ash or scrubber carry over build-up
- Latent design, fabrication and erection defects

8.3.2   Surface Corrosion, Erosion and Weathering

With regard to both erosion and corrosion, FRP liners and stacks have performed well and neither actions have been the cause of a major failure. However, maintenance repairs have to be performed occasionally in order to maintain the interior and exterior corrosion resistant surfaces. The exterior of liners and stacks that are exposed to normal environmental conditions will weather and be in need of some surface repair during the life of the liner.

8.3.3   Thermal

If a thermal failure occurs, it is probably the result of an unusual event or an operation error. FRP liners and stacks perform well within their design temperature range, but do not handle temperatures above their design range very well. Some permanent deformations may occur when exposed to excessively high temperatures.

Since FRP expands about twice as much as steel, expansion joints, lateral braces and other accessories have to accommodate these sometimes large movements. Inspectors should note the condition of these expansion points.

8.3.4   Vibration

Reinforced fiberglass structural members are inherently flexible and the designer has to allow for this. Stiffening is achieved by either large external stiffeners or internal bracing. FRP liners and stacks, or their components, have experienced damage from vibration. Therefore, during an inspection, the integrity of the stiffening has to be checked.

8.3.5 Excessive Ash or Scrubber Carry-over Build-up

Most liners, stacks, and ductwork have been designed for a certain amount of build-up. A typical value is 5 psf. The owner should be aware of this value and monitor the actual build-up so that the design values are not exceeded.

8.3.6 Latent Design, Fabrication and Erection Defects

Unfortunately, the first generation of FRP liners and stacks has experienced a number of failures due to latent design, fabrication and erection defects. The type of defects cannot be categorized into just a few areas, but there has been a relationship established between the level of quality control and assurance performed during these phases and the useful life of the liner.

If the owner or owner's representative has the QA/QC construction documentation then there is a good chance that the work was performed per the owner's and manufacturer's specification, and the likelihood of latent defects should be minimized. A detailed QA/QC discussion can be found in the ASTM D5364-93 document.

8.4 Inspection Programs

8.4.1 General

The FRP liner or stack is designed as an integral unit in which the performance of the FRP is greatly affected by the various components (i.e. support grillage, lateral restraints, insulation, etc.). Furthermore, the performance of the FRP liner or stack will affect the overall performance of the outer steel or reinforced concrete shell.

Consequently, it is of equal importance to perform a visual inspection of the FRP liner or stack, both internally and externally (from the air space), when accessible. The external visual inspection will be critical in terms of the examination of various components and the internal inspection will provide the opportunity to examine the condition of FRP sections.

In addition to visual examination, the FRP plate thickness should be monitored using an ultrasonic thickness gauge. Also, the surface hardness should be checked using the Barcol Hardness Test (ASTM D2583). The expected Barcol readings vary for each resin but in general should be between 40 and 50.

8.4.2 Class I Inspection

a) External Visual Inspection

The exterior of the FRP liner or stack shall be visually inspected from existing access points (i.e. ladders, service platforms, elevators, etc.) where available. This external visual inspection should serve to closely examine the condition of three specific areas:

- The exterior FRP liner or stack surface including all construction/fabrication structural joints and circumferential stiffeners.

- The support structures and related components; including grillage, ring girders and lateral restraints. Support grillage should be examined for deformation, over-stressing and/or general condition. The inspection should also include attachment/anchorage or support points at the reinforced concrete outer shell or skeletal frame.

- The lateral restraint system(s) should also be examined for deformation, over-stressing and/or general condition. Systems which provide for movement tolerances (i.e. tangential tie rods, spring loaded pipe hangers, etc.) should be examined carefully and the appropriate movement clearances should be noted and outlined in the report.

- All appurtenances; including items such as expansion joints, rainhood, access doors, service platforms, ladders, sample ports, etc.

b) Internal Visual Inspection

If possible, the interior of the FRP liner or stack should be visually inspected from one full height transit (drop). This internal visual inspection should serve to closely examine the condition of the FRP surface, floor, expansion joints and breeching connection.

A sample "Inspection Checklist and Records Form" is given in the appendix. This form presents a general list of items the inspector should report on. Some obvious conditions, like holes and cracks, have been omitted. Since an FRP liner or stack is an assembly of individual sections, each section should be inspected and the section elevations noted on the form.

In addition, specific attention should be given to the identification, orientation and size of any areas of distress; such as corrosion, deformed/buckled plate, cracked welds or other signs of over-stressing. The extent of surface corrosion and/or pitting should be evaluated and reported.

8.4.3 Class II Inspection

a) External Visual Inspection

The external visual inspection of the FRP liner or stack should include examination of items identified in Section 8.4.2-a, performed from existing access points (i.e. ladders, elevators, service platforms), as well as from at least one full height transit (drop) 180° from the ladder and/or elevator.

The full height transit (drop) should provide direct physical access to the external surface of the FRP and should also be orientated to include a close-up examination of the lateral restraint mechanism, as permitted.

b) Internal Visual Inspection

An internal visual inspection should include the examination of items identified in Section 8.4.2-b from four quadrants, located 90° apart. It is usually necessary to completely wash down the interior surface before a Class II Inspection is performed.

c) Ultrasonic Thickness Testing

FRP plate thickness should be obtained using an ultrasonic thickness gauge. Ultrasonic thickness measurements should be obtained on four quadrants, located 90° apart and vertically on 10 ft intervals. Additional readings should be obtained adjacent to any areas of distress.

8.4.4 Class III Inspection

A Class III inspection should include all items as outlined in a Class II inspection, as well as Non-Destructive Examination to check the integrity of critical joints and overlays. A critical joint should be defined as a highly stressed area in shear or tension where failure may result in collapse of the liner or stack.

FRP STACKS AND LINERS
INSPECTION CHECKLIST AND RECORD FORM

Stack / Flue Designation: _____

Date of Inspection: _____ Inspector's Name: _____

Ambient Conditions: _____

EXTERIOR	*OBSERVATION / COMMENT / RECORDING*
Discoloration	
Stains	
Erosion	
Corrosion	
Cracks	
Holes	
Buckling	
Blisters	
Delaminations	
Wet areas	
Deposits / Ash or Sludge Accumulation	
Sounding	
Surface Hardness	
Thickness Measurement (non-destructive)	
Thickness Measurement (destructive)	

Note: Show location of item on expanded elevation view or other appropriate media.

FRP STACKS AND LINERS
INSPECTION CHECKLIST AND RECORD FORM (continued)

Stack / Flue Designation:_____

Date of Inspection:_____ Inspector's Name:_____

Ambient Conditions:_____

INTERIOR	*OBSERVATION / COMMENT / RECORDING*
Discoloration	
Stains	
Erosion	
Corrosion	
Cracks	
Holes	
Buckling	
Blisters	
Delaminations	
Wet areas	
Deposits / Ash or Sludge Accumulation	
Sounding	
Surface Hardness	
Thickness Measurements (non-destructive)	
Thickness Measurements (destructive)	

Note: Show location of item on expanded elevation view or other appropriate media.

FRP STACKS AND LINERS
INSPECTION CHECKLIST AND RECORD FORM (continued)

Stack / Flue Designation: _____

Date of Inspection: _____ Inspector's Name: _____

Ambient Conditions: _____

APPURTENANCES	*OBSERVATION / COMMENT / RECORDING*
Base	
Supports	
Anchor Bolts	
Ladders	
Platforms	
Emissions Monitoring Ports	
Grounding	
Lightning Protection System	
Lighting	
Insulation	
Lagging	
Expansion Joints	
Access Doors	

Note: Show location of item on expanded elevation view or other appropriate media.

9.0 COATINGS AND LININGS

Because there are many different materials used in the construction of chimneys and stacks, and chimney operating conditions vary considerably, there is a wide variety of coating and lining products used to protect them. Each product is generally part of a coating or lining system that has been designed to protect a particular substrate material from specific aggressive components of the flue gas stream or the atmosphere, and each system generally requires a specific application technique.

The subject of coatings and linings is therefore complex, and it is beyond the scope of these guidelines to describe proper inspection and assessment methods for all coating and linings. Personnel familiar with the particular system employed, and the purpose for which it was designed and installed must perform the inspection. The subject has been thoroughly covered in previous publications focused specifically on the science of coating and lining design and inspection, and the reader seeking more information on the subject is referred to the following:

9.1 Coating Manuals and Standards

ASME STS-1   Steel Stack Standard -American Society of Mechanical Engineers, Three Park Avenue, New York, NY 10016-5990 www.asme.org

ASTM STP 837 Manual of Protective Linings for Flue Gas Desulfurization Systems - American Society for Testing and Materials, 100 Barr Harbor Drive, West Conshohocken, PA. www.astm.org

ASTM C 868   Standard Test Method For Chemical Resistance Of Protective linings

ASTM D 4618   Standard Specification For Design And Fabrication Of Flue Gas Desulfurization Components For Protective Lining Application

ASTM D 4619   Standard Practice For Inspection of Linings in Operating Flue Gas Desulfurization Systems

CICIND   Chimney Coatings Manual - International Committee on Industrial Chimneys – CICIND, The Secretary. 14 The Chestnuts, Beechwood Park, Hemel Hempstead, Hertfordshire, United Kingdom HP3 0DZ,

SSPC   The Inspection of Coatings And Linings: A Handbook Of Basic Practice For Inspectors' Owners, And Specifiers - Steel Structures Painting Council, 4400 Fifth Avenue, Pittsburgh, PA 15213 www.sspc.org

Other sources of information related to coatings and linings are:

NACE   The Corrosion Society (Formerly the National Association of Corrosion Engineers)
NACE International, 1440 South Creek Drive, Houston, TX 77084
www.nace.org

10.0 APPURTENANCES

10.1 Inspection of Appurtenances

Inspection of chimney appurtenances can be accomplished for the most part from the permanent access structures. However, additional temporary rigging may be required should a much closer investigation be required. Following are suggestions of areas to examine. These are only suggestions and the actual scope of each inspection should be designed to suit the specific chimney and appurtenances being inspected.

In general, the observed condition of all accessible appurtenances should be documented in any of the three classes of inspection. The actual appurtenances accessible will be determined by the class of inspection and the chimney or stack type.

Access Doors - Check door seals for degradation and signs of leakage. Insure that door hinges and latching mechanisms are operable.

Access Ladders - Inspect the condition of the attachment mechanism of the ladder brackets, Safe-T climb rail and cage. Look for damaged ladder rungs and corrosion problems in general.

Access Platforms - Examine all attachments to the chimney shell. Look for loose bolted connections or concrete anchors, corrosion problems, weld cracks, as well as corrosion problems with structural steel members, grating, and handrail. Examine grating integrity of grating clips.

Aviation Warning Lights - Check to make sure that all lights are functioning properly. If not, necessary repairs should be made immediately.

Breechings - Inspect the condition of the steel. Ultrasonic testing of the steel may be required to determine the rate of corrosion. The expansion joint should be examined for signs of leakage and the seals or packing around the opening should be looked at to access the condition.

Continuous Emissions Monitoring System (CEMS) - Check for leaks around openings (ports) in the liner where flue gas may escape and damage the concrete shell. NOTE: This inspection does not certify the operation of the CEMS. The operation of the CEMS should be checked by qualified personnel at the intervals recommended by the manufacturer or demanded by local regulations, whichever is the most frequent.

Elevators - Examine the structural connection of the rails to the shell. NOTE: This is not an elevator safety inspection. The elevator should be thoroughly inspected by a qualified elevator technician at regularly scheduled intervals to insure mechanical soundness and that all safety devices are functioning properly.

Equipment Enclosures - Check the siding for loose panels caused by missing lagging screws as a result of corrosion. Examine all structural framing connections for any flaws. Examine the condition of the painted steel.

Jib Crane - Examine the condition of all connections and general condition of the steel.

Lightning Protection/Grounding System -Examine lightning rods for damage and to insure that they are still attached to the caps. Check all grounding connections to steel structures that are attached to the chimney shell.

Liquid Collectors - Check collectors for corrosion problems that can be worsened by buildup of solids. This also affects the efficiency of the collectors, which can lead to acid moisture fallout from the chimneys.

Microwave Equipment - Examine the condition of structural steel and all connections. NOTE: This inspection does not certify the proper operation of the microwave equipment. The operation and alignment of the microwave equipment are the responsibility of appropriately trained personnel hired or employed by the owners of the antennae.

Platform/Ladder Lighting - Make note of any lights at the platforms and along ladders that are not operating.

Pressurization System - Check operating condition of fans and examine ductwork for overall condition.

Rainhood - Check for corrosion problems or areas where leaking may occur. Make sure that there is adequate space between the outside of the shell and the lip of the rainhood overhang.

Shell/Lining Caps - Examine the condition of the steel caps and the bolted connections between segments. Ultrasonic testing of the caps may be required to determine steel thickness remaining. Also insure that the bolts are in place and tight.

Turning Vanes - Check for connection problems as well as any damage to vanes caused by corrosion and/or external forces.

11.0 GLOSSARY OF TERMS

Access Doors - Doors located at the top, bottom or intermediate levels, which provide access to the interior of the shell and/or liner.

Access Ladders - Ladders attached to the shell or liner between platforms. Ladders may be caged or equipped with safety climb devices.

Access Platforms - Platforms at various elevations of chimneys/stacks that are attached externally or internally to the shell. Platforms may provide up to 360-degree access or be solely used as a rest platform during climbing.

Annulus - The space between the outer shell or weathershield of a chimney, and the inner liner.

Aviation Warning Lights - Lights (Strobe, blinking or steady) attached to platforms or built into the shell to warn approaching aircraft.

Breeching - Frame in an opening in the outer sheath and liner through which the flue duct is installed.

CEMS - Continuous Emissions Monitoring System required by the EPA for monitoring quality of exhaust gas.

Drop - Full-height access for inspection personnel to a chimney, stack, or liner. May be provided by self-climbing scaffold platform, swing-stage, bosun's chair, or rappelling equipment, depending on the type of inspection.

Elevators - Typically rack and pinion type elevators attached internally or externally to the shell. These elevators may or may not provide access to all platform levels and are mainly used for access to instrumentation (CEMS) and inspection platforms.

Equipment Enclosures - Used to protect installed instrumentation equipment from weather.

FGD - Flue Gas Desulfurization system.

Jib Crane - Cantilever crane located at platforms used to hoist equipment from the ground using cables and cable blocks.

94   CHIMNEY AND STACK INSPECTION GUIDELINES

Lightning Protection/Grounding System (LPS) - Includes lightning rods attached to the rainhood or caps to protect against lightning strikes directly to the chimney. Grounding cables are also attached to platforms and cables run vertically from top to bottom and circumferentially at various elevations. LPS is generally not required on steel stacks.

Liner -   A structure that prevents the exhaust gases from coming into contact with the shell of a chimney. Liner types are:

Independent: A free standing column that is constructed and supported independently of the outer shell.

Corbelled: A sectional liner, where each interlocking section is supported on a shelf, or corbel, constructed on the inner face of the shell.

Suspended: A lightweight liner, usually steel or fiberglass, which is suspended from the outer shell.

Contiguous: A liner that is constructed in contact with the inner face of the outer shell, and is usually supported by it.

Liquid Collectors - Collection system which is used to divert moisture inside the chimney liner in such a way that it will not be re-entrained into the gas path and flow out the top of the chimney.

Microwave Equipment - Structural frame and antenna (e) attached to the chimney shell for microwave radio communications.

Mass Damper - Device to suppress motion caused by crosswinds.

NDT -   Nondestructive testing. Physical testing that causes no significant damage to, or disfiguration of, the structure.

Particulates -   Fine particles that are carried in the flue gas stream.

Platform Lighting - General lighting, typically found either on the outside or in the annulus of a chimney.

Ports -   Openings in the shell and/or liner for the installation of monitoring instruments.

Precipitator -   Dry system for removing particulates from the flue gas stream before discharge through the chimney.

Pressurization System - Mechanical system comprised of ducts and fans that is used to pressurize the annulus of chimneys. Generally used at plants that utilize wet scrubbers in order to insure that wet gases do not get into the annulus and attack the outside shell.

Profile - A graph or drawing representing conditions along a vertical line up the chimney, stack, or liner.

Rainhood - Typically attached to the top of a chimney (steel, concrete or FRP) to provide weather protection for the annulus.

Scrubber - Wet system for removing particulate matter from the flue gas stream before discharge through the chimney.

Shell - The outer part of a chimney. On multi-flued stacks, also known as the WINDSHIELD or WEATHERSHIELD.

Shell/Lining Caps - Steel or cast iron caps that protect the top of the liner and shell on concrete or brick chimneys.

Shell Opening Reinforcement - Frame at the opening in the shell through which the breeching passes.

Strakes - Fins installed on the exterior of a stack to break up critical wind vortices.

Thimble - Sheet metal flue duct across the annular space of a chimney.

Turning Vanes - Vanes constructed from steel or FRP, which are used to divert flue gas vertically inside the chimney or stack. Typically installed where two or more flues enter a common chimney.

Note: Where trade names of equipment or test methods are mentioned in this guide, they are clearly identified by italic letters. Use of trade names does not imply any endorsement or preference for the named item. The name is used only to provide a well-known example of a generic type of equipment or test method, and avoid possible confusion with similarly named but different products or services.

APPENDIX A - SAMPLE INSPECTION REPORT SPECIFICATION

The following is a sample report structure specification modeled on one issued by Florida Power and Light Company. It is not intended to be a standard specification document, but lists those instructions and options which it would be considered reasonable to include in a specification. Actual instructions and requirements should be varied according to the needs of the owner, and the type and age of the structure being inspected:

(1) A bound report shall be prepared, structured, and submitted in the following manner:

    Section 1.0 - Executive Summary
    Section 2.0 - Introduction
    Section 3.0 - Inspection Methodology and Procedures
    Section 4.0 - Structure Description
    Section 5.0 - Inspection Findings and Engineering Review
    Section 6.0 - Condition Assessment and Repair Recommendations
    Section 7.0 - Budget Cost Estimates for Repairs
    Section 8.0 - Appendices
        Appendix A - Photographs
        Appendix B - Drawings and Sketches
        Appendix C - Field Inspection Data Forms
        Appendix D - Laboratory Test Reports

(2) Section 5.0 must include all the information obtained from the inspections and laboratory tests so that it is possible to issue repair recommendations on the basis of this information. This must at least include the following:

- A general impression of the condition of the component
- An interpretation of findings from the inspections and laboratory tests
- A determination of the nature and extent of the damage
- An estimate of the rate of aging or degradation of the damaged parts

(3) Sections 6.0 and 7.0 will depend on the findings given in Section 5.0. At a minimum these sections will include discussion of:

- The probable cause of the damage
- Estimated rate of degradation and residual service life of the damaged part
- The residual life required
- Possible repairs and estimated service life extension
- Short-term and long-term repair costs
- Repair plan/schedule
- Repair recommendations with reasons

(4) The inspection findings, analysis, and repair recommendations shall be performed and/or reviewed by a registered professional engineer. Calculations shall be performed on an as-needed basis to determine inspection finding results with regard to the structural integrity of the chimney

(5) Three copies of a preliminary report shall be submitted to the owner for review and comment. After receiving the comments, the inspection firm shall act on or incorporate the comments and submit one original and two full-color copies of the final report.

Samples of inspection checklists to document work done and observations made are given in Appendices A and B of this guide.

APPENDIX B – SAMPLE CHECKLISTS AND FORMS

Inspection Checklist Example No. 1

TABLE OF CONTENTS

DOCUMENT REVISIONS ii

PAGE REVISION CONTROL iii

1.0 SCOPE... 1

2.0 INTERNAL INSPECTION 1
    2.1 General 1
    2.2 Brick Liners – Free Standing 1
    2.3 Brick Liners – Corbel Supported 1
    2.4 Liner Displacement Survey 2

3.0 ANNULAR SPACE INSPECTION 2
    3.1 General 2

4.0 EXTERNAL INSPECTION 2
    4.1 General 2
    4.2 Concrete Cores 3
    4.3 Appurtenances 3

5.0 PHOTOGRAPHS 3

6.0 REPORTS.. 4
    6.1 General 4
    6.2 Report Structure 4
    6.3 Engineering Review.. 5

7.0 SUBMITTALS 5
    7.1 Field and Report Data 5

DOCUMENT REVISIONS

REVISIONS **DATE**

PAGE REVISION CONTROL

Page	Date
1	
2	
3	
4	
5	
6	
7	
8	
9	

SAMPLE CHIMNEY INSPECTION SPECIFICATION

1.0 SCOPE

The purpose of this specification is to provide generic criteria for performing chimney assessment inspections. Inspections shall be performed on a recommended two-year basis. The implementation of scheduled inspections in accordance with these criteria should be utilized as a planning tool to provide standardized chimney report formats, identify and upgrade deficiencies, eliminate unscheduled repairs and outages, and reduce overall maintenance costs by minimizing out of scope work during maintenance projects. In addition, the program should provide a systematic approach toward the assessment and evaluation of present conditions, allowing the opportunity to schedule repairs that are timely and well planned.

2.0 INTERNAL INSPECTION

2.1 General

The inspection of the chimney liner shall include an examination and evaluation of the lining materials and any baffles, access doors, flue connections or other appurtenances accessible from the interior of the chimney. Accumulated fly ash or rust shall be removed, as necessary, to properly evaluate the condition of items or features so covered.

1. Inspection procedures shall involve the installation of rigging which will permit direct contact with lining walls at a minimum of four locations, 90° apart.

2. For reporting purposes, the liner shall be divided into four quadrants located 90° apart and vertically divided into equal sections, with a minimum spacing of 25 ft and maximum spacing of 50 ft.

3. The findings within each section shall be identified and reported, with specific attention being given to the identification, orientation, and size and length of any cracks, spalling or any other areas of distress.

2.2 Brick Liners – Free Standing

For those chimneys with free-standing brick liners, the integrity of the lining shall be evaluated on a visual basis, supplemented by routine probing to determine hardness, soundness and/or general condition of the brickwork.

2.3 Brick Liners – Corbel Supported

For those chimneys with corbel supported brick liners, the integrity of the lining shall be evaluated on a visual basis, supplemented by routine probing to determine hardness, soundness, and/or general condition of the brickwork.

The location of all in-place and missing (or cracked) corbel block shall be documented per Section 6.0. The condition of corbel packing shall be identified as possible.

2.4 Liner Displacement Survey

A liner displacement survey shall be performed utilizing electronic survey equipment. Measurements identifying the extent of liner displacement (if any) shall be provided for each lining section, as described in Section 2.1.3.

3.0 ANNULAR SPACE INSPECTION

3.1 General

For those chimneys with an annular space accessible with a bosun's chair, the annular space shall be rigged to facilitate the inspection of the annular space at three locations, throughout the entire height. The initial drop shall be located on the wall between the flue openings and the other two equally spaced circumferencially from the first.

Where the annular space is too narrow for visual inspection, video examinations shall be provided continuously on a minimum of three locations, as specified above. The camera shall be raised at a rate of 10 ft/min.

1. The annular space inspection shall provide for a visual examination of the reinforced concrete shell, brick liner, buckstay system, flue openings, pilasters and breeching ductwork. <u>A special effort shall be made to inspect each and every liner reinforcing band, and associates bolts.</u> Accumulated fly ash or rust shall be removed as necessary to properly evaluate the condition of the items or features so covered. Photographic documentation shall be provided as described in Section 5.0 and inspection finding requirements as detailed in Section 6.0.

2. The findings shall be identified and reported with specific attention being given to the identification, orientation and size of any cracks, spalling or other areas of distress.

4.0 EXTERNAL INSPECTION

4.1 General

The exterior surface of the reinforced concrete outer shell shall be rigged to facilitate the inspection at four locations 90° from each other the entire length of the shell.

1. The exterior concrete surface shall be evaluated on a visual basis, supplemented by routine probing to determine hardness, soundness, and/or general condition of the concrete.

2. The findings shall be identified and reported with specific attention being given to the identification, orientation and size and length of any cracks, spalling or other areas of distress.

4.2 Concrete Cores (Not to be performed on a routine basis; only upon request by the owner)

For those chimneys with corbel supported brick liners, two 4" diameter concrete core samples shall be removed from elevations corresponding to a maximum 12" above the top of shelf corbel projections, formed in the reinforced concrete outer shell.

The concrete cores shall be removed in accordance with ASTM C-42 "Drilled Cores and Sawed Beams of Concrete Obtaining and Testing". The cores and test results (if required) shall be submitted with the inspection report. An optional price shall be included for the cores to be tested by the contractor as follows:

Petrographic study in accordance with ASTM C-856 "Petrographic Examination on Hardened Concrete" and other tests as necessary to obtain:

a) A description of the cementitious matrix, including qualitative determination of the binder, degree of hydration, degree of carbonation, and degree of sulfate attack.

b) An evaluation of the matrix; including the evaluation of water-cement ratio, paste content, cracking, degree of consolidation, and degree of degradation.

c) An evaluation of the aggregate; including gradation, maximum aggregate size, type of aggregate, hardness, degradation of the aggregate, and possible chemical interaction between the matrix and the aggregate.

d) An evaluation of the reinforcing steel; including size, description, and degree of corrosion.

e) Compressive strength of the concrete.

4.3 Appurtenances

Appurtenances shall consist of but not be limited to cleanout door, ladder, platforms, reinforcing bands, obstruction lights and conduits, lightening protection system and gutter systems.

All appurtenances shall be inspected visually and the findings identified and reported with specific attention being given to any structural defects, corrosion, or other areas of distress. Photographic documentation as described in Section 5.0 and inspection finding requirements as described in Section 6.0.

5.0 PHOTOGRAPHS

1. To document the conditions found to the existing chimney and to facilitate the evaluation of these conditions, the contractor shall record all deficiencies and questionable features using 35 mm color photographs, taken in sufficient quantity and detail to adequately define the extent of all deficiencies.

2. In addition to this requirement, the following areas shall also be photographed to be included in the report:

   a) Interior photographs shall be taken on a minimum of two sides, at 50 ft intervals. Additional photographs of all detail areas, baffles, access doors, flue connection, pilasters, buckstays, etc. shall be taken in sufficient quantity to identify the general condition.

   b) Exterior photographs shall be taken on a minimum of 50 ft elevations, at the four locations specified in Section 4.1.

   c) Appurtenance photographs shall be taken of all items in sufficient quantity to identify the general condition.

6.0 REPORTS

6.1 General

The report shall log all items inspected along with a description of the findings. The report shall include detailed sketches and photographs to completely describe the condition as found. The report shall also contain the following sections; Inspection Methodology and Procedures, Structural Description, Inspection Findings, Repair Recommendations and associated budget cost estimates.

6.2 Report Structure

1. The bound report shall be structured in the following manner:

   Section 1.0 -- Executive Summary
   Section 2.0 – Introduction
   Section 3.0 – Inspection Methodology and Procedures
   Section 4.0 – Structure Description
   Section 5.0 – Inspection Findings and Engineering Review
   Section 6.0 – General Discussion and Repair Recommendations
   Section 7.0 – Budget Cost Estimates
   Section 8.0 – Appendices
       Appendix A – Photographs
       Appendix B – Drawings and Sketches
       Appendix C – Field Inspection Data Forms (Sample attached)
       Appendix D – Laboratory Test Reports

2. Section 5.0 must include all the information obtained from the investigations and laboratory tests so that it is possible to issue repair recommendations on the basis of this information. This must at least include the following:

   - The general impression of the condition of a component.
   - The interpretation of findings from investigations and laboratory tests.
   - Determining the nature and extent of the damage.
   - Determining the rate of aging of the damaged parts.

3. Sections 6.0 and 7.0 are the result of the inspection report in which recommendations are made with regard to the measures to be taken. This must contain the following:

   - The probable cause of damage.
   - The anticipated development of damage and the residual life required of the part which is damaged.
   - The residual life required.
   - Possible repairs and anticipated life.
   - Short-term and long-term costs.
   - Planning.
   - Recommendations with reasons.

6.3 Engineering Review

The inspection findings and repair recommendations shall be reviewed by a professional engineer registered in the State of Florida. The Engineer's name shall be submitted with the contract proposal. Calculations shall be performed on an as-needed basis to determine inspection finding results on the structural integrity of the chimney.

7.0 SUBMITTALS

1. Prior to the start of work, the contractor shall submit formatted blank inspection data sketches to the owner for approval.

2. Sketches shall be developed in detail so as to completely describe all features of the chimney design and various components.

3. Three copies of a preliminary report shall be submitted to the owner for review and comment. After receiving the comments, the contractor shall incorporate the comments and submit four copies of the final report. The original and two (2) color copies, containing photographic/xerox color prints and one (1) xerox copy shall be submitted.

SAMPLE FIELD INSPECTION DATA FORM – Format No. 1

INSPECTION REPORT

Inspection Number :
Date :
Name of Inspector :
Inspection Company:

1 SHELL **Remarks**
   1.1 Uniform degradation
   1.2 Local degradation
   1.3 Wear
   1.4 Cracks
   1.5 Joints
   1.6 Coating
   1.7 Sweating/color changes
   1.8 Sundries

2 TOP
   2.1 Coverage
   2.2 Shell
   2.3 Lining
   2.4 Insulation
   2.5 Coating
   2.6 Lightning protection
   2.7 Top platform
   2.8 Cage ladder
   2.9 Fastening structures
   2.10 Sundries

3 INSIDE OF LINING
3.1 Brickwork
   3.1.1 Uniform degradation
   3.1.2 Local degradation
   3.1.3 Cracks
   3.1.4 Joints
   3.1.5 Expansion joints (inc. posterior parts)
   3.1.6 Cakes of ash

3.1.7 Sweating/color changes
   3.1.8 Sundries
3.2 Steel
   3.2.1 Uniform degradation
   3.2.2 Local degradation
   3.2.3 Uniform rust deposit
   3.2.4 Local rust deposit
   3.2.5 Cracks
   3.2.6 Welded seams
   3.2.7 Expansion joints
   3.2.8 Wall thickness
   3.2.9 Coating
   3.2.10 Fastening structures
   3.2.11 Cakes of ash
   3.2.12 Condensate
   3.2.13 Sundries

4 ACCESSIBLE AIR SPACE
4.1 Inside Shell
   4.1.1 Uniform degradation
   4.1.2 Local degradation
   4.1.3 Cracks
   4.1.4 Joints
   4.1.5 Condensate
   4.1.6 Drains
   4.1.7 Movable parts (doors, manholes, etc.)
   4.1.8 Ventilation provisions
   4.1.9 Sundries
4.2 Outside Brickwork Lining
   4.2.1 Uniform degradation
   4.2.2 Local degradation
   4.2.3 Cracks
   4.2.4 Joints
   4.2.5 Expansion joints (inc. Posterior parts)
   4.2.6 Supporting structure
   4.2.7 Sweating/color changes
   4.2.8 Condensate
   4.2.9 Insulation
   4.2.10 Sundries

CHIMNEY AND STACK INSPECTION GUIDELINES

4.3 Outside Steel Lining
    4.3.1 Uniform rust deposit
    4.3.2 Local rust deposit
    4.3.3 Cracks
    4.3.4 Welded seams
    4.3.5 Expansion joints
    4.3.6 Wall thickness
    4.3.7 Coating
    4.3.8 Fastening structures
    4.3.9 Condensate
    4.3.10 Insulation
    4.3.11 Sundries

5 PLATFORMS

5.1 Degradation (uniform, local)
    5.1.1 Upper side (steel) structure
    5.1.2 Lower side (steel) structure
    5.1.3 Grating
    5.1.4 Handrail
    5.1.5 Stays and supports
5.2 Rust deposit (uniform, local)
    5.2.1 Upper side (steel) structure
    5.2.2 Lower side (steel) structure
    5.2.3 Grating
    5.2.4 Handrail
    5.2.5 Stays and supports
5.3 Cracks
5.4 Welded seams
5.5 Coating
5.6 Fastening structures
5.7 Slope of floors and drains
5.8 Sundries

6 FLUE GAS INLET

6.1 Expansion provisions at inlet
6.2 Support structure flue gas ducts
6.3 Compensator
6.4 Condensate and ash funnel
6.5 Condensate collecting spout

6.6 Drains
6.7 Sundries

7 **EQUIPMENT**
7.1 Cage ladders
7.2 Climb safety provisions
7.3 Connections hooks/fastener
7.4 Tapes and closures
7.5 Coating
7.6 Lightning protection leads
7.7 Aircraft warning lights
7.8 Inside illumination
7.9 Measuring arrangements
7.10 Sundries

CHIMNEY AND STACK INSPECTION GUIDELINES 111

SAMPLE FIELD INSPECTION DATA FORM – Format No. 2

INSPECTION REPORT

Number: Date:
Name of Inspector:
Inspection Company:

CHECKLIST

		No Remarks	Explanation Number
1	**SHELL**	O	
1.1	Uniform degradation	O	
1.2	Local Degradation	O	
1.3	Wear	O	
1.4	Cracks	O	
1.5	Joints	O	
1.6	Coating	O	
1.7	Sweating/color changes	O	
1.8	Sundries	O	
2	**TOP**	O	
2.1	Coverage	O	
2.2	Shell	O	
2.3	Lining	O	
2.4	Insulation	O	
2.5	Coating	O	
2.6	Lighting protection	O	
2.7	Top platform	O	
2.8	Cageladder	O	
2.9	Fastening structures	O	
2.10	Sundries	O	
3	**INSIDE OF LINING** (flue gas side)	O	
3.1	Brickwork	O	
3.1.1	Uniform degradation	O	
3.1.2	Local degradation	O	
3.1.3	Cracks	O	
3.1.4	Joints	O	
3.1.5	Expansion joints (incl. posterior parts)	O	
3.1.6	Cakes of ash	O	
3.1.7	Sweating/color changes	O	

3.1.8	Sundries	O
3.2	Steel	O
3.2.1	Uniform degradation	O
3.2.2	Local degradation	O
3.2.3	Uniform rust deposit	O
3.2.4	Local rust deposit	O
	INSIDE OF LINING (Continued)	
3.2.5	Cracks	O
3.2.6	Welded seams	O
3.2.7	Expansion joints	O
3.2.8	Wall thickness	O
3.2.9	Coating	O
3.2.10	Fastening structures	O
3.2.11	Cakes of ash	O
3.2.12	Condensate	O
3.2.13	Sundries	O
4	**ACCESSIBLE AIR SPACE**	O
4.1	Inside shell	O
4.1.1	Uniform degradation	O
4.1.2	Local degradation	O
4.1.3	Cracks	O
4.1.4	Joints	O
4.1.5	Condensate	O
4.1.6	Drains	O
4.1.7	Movable parts (doors, manholes, etc.)	O
4.1.8	Ventilation provisions	O
4.1.9	Sundries	O
4.2	Outside brickwork lining	O
4.2.1	Uniform degradation	O
4.2.2	Local degradation	O
4.2.3	Cracks	O
4.2.4	Joints	O

CHIMNEY AND STACK INSPECTION GUIDELINES

4.2.5	Expansion joints (incl. Posterior parts)	O
4.2.6	Supporting structure	O
4.2.7	Sweating/color changes	O
4.2.8	Condensate	O
4.2.9	Insulation	O
4.2.10	Sundries	O
4.3	Outside steel lining	O
4.3.1	Uniform rust deposit	O
4.3.2	Local rust deposit	O
4.3.3	Cracks	O
4.3.4	Welded seams	O
4.3.5	Expansion joints	O
4.3.6	Wall thickness	O
4.3.7	Coating	O
4.3.8	Fastening structures	O
4.3.9	Condensate	O
4.3.10	Insulation	O
4.3.11	Sundries	O
5	**PLATFORMS**	O
5.1	Degradation (uniform, local)	O
5.1.1	Upper side (steel) structure	O
5.1.2	Lower side (steel) structure	O
5.1.3	Grating	O
5.1.4	Handrail	O
5.1.5	Stays and supports	O
5.2	Rust deposit (uniform, local)	O
5.2.1	Upper side (steel) structure	O
5.2.2	Lower side (steel) structure	O
5.2.3	Grating	O

114   CHIMNEY AND STACK INSPECTION GUIDELINES

5.2.4	Handrail	O	_____
5.2.5	Stays and supports	O	_____
5.3	Cracks	O	_____
5.4	Welded seams	O	_____
5.5	Coating	O	_____
5.6	Fastening structures	O	_____
5.7	Slope of floors	O	_____
5.8	Sundries	O	_____
6	**FLUE GAS INLET**	O	_____
6.1	Expansion provisions at inlet	O	_____
6.2	Supporting structure flue gas ducts	O	_____
6.3	Compensator	O	_____
6.4	Condensate and ash funnel	O	_____
6.5	Condensate collecting spout	O	_____
6.6	Drains	O	_____
6.7	Sundries	O	_____
7	**EQUIPMENT**	O	_____
7.1	Cage ladders	O	_____
7.2	Climb safety provisions	O	_____
7.3	Connections hooks/fastener	O	_____
7.4	Tapes and closures	O	_____
7.5	Coating	O	_____
7.6	Lightning protection leads	O	_____
7.7	Aircraft warnings lights	O	_____
7.8	Inside illumination	O	_____
7.9	Measuring arrangements	O	_____
7.10	Sundries	O	_____

INSPECTION REPORT

Number: Date:

Name of Inspector:

Company of Inspector:

Explanation Number

Item:

Remarks:

Item:

Remarks:

INSPECTION REPORT

Number:

Name of Inspector:

Company of Inspector:

Conclusions
Explanation:

This must include all the information obtained from the investigations and laboratory tests so that it is possible to issue a maintenance advice on the basis of this information. This must at least include the following:

> The general impression of the condition of a component.
> Interpretation of conclusions from investigations and laboratory tests.
> Determining the nature and extent of the damage (See Figure Annex 1-1)
> Determining the rate of aging of the damaged parts

Maintenance Advice

Number of Inspection Report: Date:

Name of Consultant:

Company of Consultant:

Advice

Explanation:

The maintenance advice is the result of the inspection report in which recommendations are made with regard to the measures to be taken. This must contain at least the following:

- The probable cause of damage.
- The anticipated development of damage and the residual life required of the part which is damaged.
- The residual life required.
- Possible repairs and anticipated life.
- Short-term and long-term costs.
- Planning
- Recommendation with reasons.

REPAIR REPORT

Number of Inspection Report: Date:

Name of Inspector/Supervisor:

Company of Inspector/Supervisor:

REPORT

Explanation:

The repair report is a report on the repair work carried out. This must at least include the following:

- Repair methods
- Extent of repair work
- Quality of the result
- Conditions during the work
- Costs of the repair work

Appendix C - Example of Developed Plan of a Concrete Utility Chimney

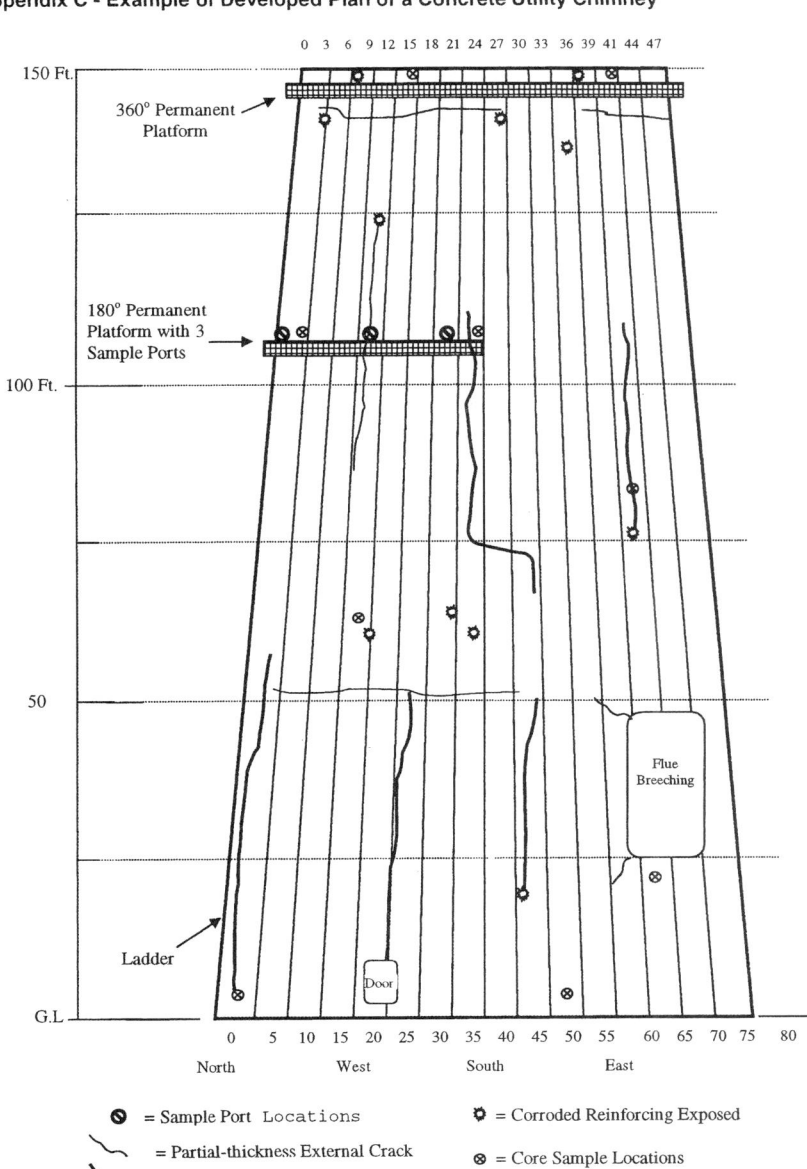

Bibliography and Standards References

Standardization Agencies or Professional Bodies referred to in these guidelines.

ASME   American Society of Mechanical Engineers
Three Park Avenue
New York, NY 10016-5990   www.asme.org

ASTM   American Society for Testing hand Materials
100 Barr Harbor Drive,
West Conshohocken, PA 19428.   www.astm.org

CICIND   International Committee on Industrial Chimneys
The Secretary, CICIND.
14 The Chestnuts,
Beechwood Park, Hemel Hempstead,
Hertfordshire, United Kingdom HP3 0DZ   www.cicind.org

EPRI   Electric Power Research Institute
3412 Hillview Avenue
Palo Alto, CA 94304   www.epri.com

NACE   The Corrosion Society (Formerly National Association of Corrosion Engineers)
NACE International
1440 South Creek Drive
Houston, TX 77084   www.nace.org

SSPC   Steel Structures Painting Council
4400 Fifth Avenue
Pittsburgh, PA 15213   www.sspc.org

Manuals, Codes of Practice and other References

ASTM STP 837 Manual of Protective Linings for Flue Gas Desulfurization Systems.

CICIND   Chimney Coatings Manual

EPRI   Report No. TR-101654 - "Guidelines for the Use of Fiberglass-Reinforced Plastic in Utility Flue Gas Desulfurization Systems

SSPC   The Inspection of Coatings and Linings: A Handbook Of Basic Practice For Inspectors' Owners, And Specifiers

Standards Referred To In These Guidelines

AASHTO T260 Standard Method of Sampling and Testing for Total Chloride Ion in Concrete and Concrete Raw Materials

ASME STS-1 Steel Stack Standard

ASTM C 62 Standard Specification for Building Brick.

ASTM C 67 Specification for Sampling and Testing Brick and Structural Clay Tile

ASTM C 90 Specification for Load-bearing Concrete Masonry Units

ASTM C 91 Standard Specifications for Masonry Cement

ASTM C 126 Specification for Ceramic Glazed Structural Clay Facing Tile, Facing Brick and Solid Masonry Units

ASTM C 150 Specifications for Portland Cement

ASTM C 597 Standard Test Method for Pulse Velocity Through Concrete

ASTM C 652 Standard Specifications for Hollow Brick

ASTM C 803 Penetration Resistance of Hardened Concrete.

ASTM C 805 Rebound Number of Hardened Concrete

ASTM C 868 Standard Test Method For Chemical Resistance Of Protective linings

ASTM C 876 Standard Test Method for Half-Cell Potentials of Uncoated Reinforcing Steel in Concrete.

ASTM C 1298 Guide for Design and Construction of Brick Liners for Industrial Chimneys

ASTM C 1383 Standard Test Method for Measuring the P-Wave Speed and the Thickness of Concrete Plates Using the Impact-Echo Method.

ASTM D 4618 Standard Specification For Design And Fabrication Of Flue Gas Desulfurization Components For Protective Lining Application

ASTM D 4619 Standard Practice For Inspection of Linings in Operating Flue Gas Desulfurization Systems

ASTM D 5364 Standard Guide for Design, Fabrication, and Erection of Fiberglass Reinforced Plastic Chimney Liners with Coal-Fired Units

ASTM E 797 Practice for Measuring Thickness by Manual Ultrasonic Pulse-Echo Contact Method

INDEX

AASHTO 18
AASHTO T260 – Standard Method of Sampling and Testing for Total Chloride Ion in Concrete and Concrete Raw Materials 27, 119
Access doors 38, 40
Access Doors 88, 91, 93
Access Ladders 91, 93
Access Platforms 91, 93
ACI 18, 34
ACI 201.1R Guide for Making a Condition Survey of Concrete in Service 37
ACI 207.3R Practices for Evaluation of Concrete in Existing Massive Structures for Service Conditions .. 37
ACI 228.2R Nondestructive Testing of Concrete 37
ACI 307 - Design and Construction of Reinforced Concrete Chimneys 37
acid attack 6, 11-12, 68
acid condensation 11-12, 47, 66, 67
acid dew point 73, 75
acid-resistant brick 67
air flow 6
air terminals 38, 40, 42, 48, 52
alkali aggregate reaction 21
Alloy cladding 55
ambient light conditions 14
American Association of State Highway Officials 18
American Concrete Institute 18
American Society for Testing and Materials 18
Anchor Bolts 58-59, 63, 88
Anchor Chairs 59
annular space . 5, 13, 15-17, 68, 69, 76, 80, 95, 102
Annulus 39, 40, 44, 93
Annulus pressurization system 39, 40
anodic 36
Antennae 48
Antimony Trioxide 81

appurtenances .. 5, 7, 17, 31, 55, 56, 59, 75-76, 82, 84, 91, 101, 104
asbestos 5
ash build-up 15
ASME STS-1 Steel Stack Standard 89, 119
ASTM 18, 46, 66, 73, 78, 83, 103, 118, 119, 120
ASTM C 62, Standard Specification for Building Brick 46
ASTM C 652, \Standard Specifications for Hollow Brick 46
ASTM C 67; \Specification for Sampling and Testing Brick and Structural Clay Tile 46
ASTM C1383 – Standard Test Method for Measuring the P-Wave Speed and the Thickness of Concrete 24
ASTM C597 – Standard Test Method for Pulse Velocity Through Concrete 21
ASTM C803 - Penetration Resistance of Hardened Concrete 21
ASTM C805 - Rebound Number of Hardened Concrete 20
ASTM C856 – Standard Practice for Petrographic Examination of Hardened Concrete 28
ASTM C876 – Standard Test Method for Half-Cell Potentials of Uncoated Reinforcing Steel in Concrete. 22
ASTM D 4618 Standard Specification For Design And Fabrication Of Flue Gas Desulfurization Components For Protective Lining Application 89
ASTM D 4619 Standard Practice For Inspection of Linings in Operating Flue Gas Desulfurization Systems . 89
ASTM D2583 – Standard Test Method for Indentation Hardness of Rigid Plastics by Means of a Barcol Impressor 29

ASTM E164 – Standard Practice for Ultrasonic Contact Examination of Weldments 25
ASTM E23 - Standard Test Methods for Notched Bar Impact Testing of Metallic Materials 28
ASTM E797 - Practice for Measuring Thickness by Manual Ultrasonic Pulse-Echo Contact Method 25
ASTM STP 837 Manual of Protective Linings for Flue Gas Desulfurization Systems 89, 118
Aviation Warning Lights 91, 93
Aviation Warning Obstruction Lighting 38, 40
Aviation Warning Obstruction Lighting 56
Aviation warning paint 38
Avongard\ crack monitor 24
Baffle walls 50
Barcol Hardness Test 29, 83
base drains 68
Base Plate 59, 63
Base Supported Liner 71
baseline data 6, 30
beam pocket 38
bellows .... See Expansion joints
binoculars 5, 16
Booster fans 66
Breeching ... 38, 40, 42, 43, 48, 50, 52, 54, 59, 63, 93
Breeching duct ... 38, 40, 42, 48, 50, 52
Breechings 79, 91
brick liner systems 67
Brick liners 66, 67, 68
brick shells 45, 46, 67
brittleness of metals ..... See ASTM E23
buckled liners 18
Buckling 58, 86, 87
building brick 45
calcium hydroxide See Portland cement
carbonation 19, 20, 28, 37, 103
Carbonation depth testing 41
cathodic 36
catwalks 5

CEMS 91, 93
Charpy Test .... See ASTM E23
chemical analysis 6, 41, 51, 70, 77
chemical attack 3, 27-28, 66, 67
chimney arrangements 4
Chimney cap .... See rainhood
Chimney Coatings Manual 89, 118
chimney development sketch 39, 49, 50
chimney maintenance 1
Chimneys and stacks 1
Chimneys of Special Importance 7
chloride content 23
Chloride content ... See AASHTO T260
chlorides 23, 28
CICIND 89, 118
Class I inspection 5, 6, 58
Class II inspection 6, 41, 77, 85
Coating .. 43, 60, 61, 89, 107, 108, 109, 110, 111, 112, 113, 114
Common Brick .... See Building Brick
concrete chimney 34
concrete quality 20, 21
Condition assessment 3
construction . 1, 3, 4, 15, 16, 18, 21, 31, 34, 35-40, 45-47, 67, 74, 79, 83, 84, 89
construction, operation, and maintenance 3
Contiguous wall liner 66
corbel 7, 11, 16, 67, 68, 70, 94, 102, 103
Corbel 50, 66, 70, 98, 102
core samples 6, 103
corrosion 5-7, 11-13, 16, 19, 20, 22, 23, 25, 28, 36-41, 47-49, 55-61, 72-78, 81-84, 91, 92, 103, 104
Corrosion 68
Corrosion Protection 55
corrosion susceptibility 6
Counter Weight Systems 80
covermeter 20
crack growth 24
crack monitoring 6, 51
crack movement .... See crack growth

cracks 5, 11, 16, 18, 24, 26, 35-40, 47-50, 58-60, 66, 68, 70, 84, 91, 101-103
database.................. 1, 7, 14-18, 30, 32
degradation........... 1, 14-16, 28, 30, 32
Degradation of Bricks....................... 68
Degradation of Flue Ducts 68
Degradation of Mortar 67
delamination 21, 23, 35
Demec gage 24
depth of cover concrete................... 20
design of chimneys 4
Differential Expansion..................... 58
discoloration 5, 14, 16
Discoloring .... See Staining
distress..... 3, 16, 17, 21, 76, 77, 84, 85, 101-104
Documentation 17
downleads................ 38, 40, 42, 48, 52
Drain .... See False Bottom Drain
Drop 93
drop locations 15, 16, 17
ductwork 15, 39, 40, 79, 80, 83, 92, 102
duty cycle 37
duty cycling 3
dynamic monitoring 4, 29, 30
dynamic stiffness........................ 23, 24
earthquake 3, 29, 30, 34
Efflorescence 16
elastomeric coatings 24
Elastomeric coatings 55
Electric Power Reseacrh Institute..... 78
elevated floors 68, 69
Elevator 39, 41
Elevators 91, 93
emissions reduction systems 7, 35
Environmental factors 15
EPRI Report No. TR-101654 78
Equipment Enclosures 92, 93
estimating concrete strength............. 20
Exit cone 56
Expansion Joints 59, 62, 79, 80, 88
explosion 12, 29, 58
Explosion... 6

exposed rebar 39, 48, 49
exterior.5, 6, 15, 26, 27, 37, 38-40, 42, 45, 47-49, 67, 69, 73, 75, 78-82, 84, 95, 103
extremes of heat and cold................... 3
falling hazard 5
False Bottom 60, 63, 64
False Bottom Drain......................... 59
Fatigue Cracks................................ 58
FGD 73, 93
fiber-reinforced plastic 3
Fire... 6
flame spread ratio. 81
Floating/Toggle Liner 71
Flooding 6
Flow straightening plates 50
flue breeching 15, 16, 17, 68
flue gas booster fans 7
flue gas desulfurization 73
flue gas flow velocity 35
flue gas leakage 38, 48, 66, 75
flue gas scrubbers 7
flue gas temperature................... 34, 35
flue liners 6
Flue Supports 60
fly ash.................... 15, 16, 17, 101, 102
flyash sludge 68
forced draft systems 35
freeze-thaw cycles........................... 37
FRP ..3, 5, 6, 12, 14, 28, 29, 57, 78-88, 95
FRP liners....................................... 82
fuel switching................................... 3
full-height inspection........................ 5
Galvanizing.................................... 55
gas dissipation 66
gas ejection velocity....................... 66
Gas sampling ports 48, 50
generating units 7
glazed brick 67
GPR 26
Ground Penetrating Radar ..... See GPR
gypsum..................................... 35, 47
Half-cell Potential 22
hazardous materials................. 5, 33

Health Hazards 5
Heat Transfer Control 56
half-cell potential 22, 23, 41
high operating temperatures 3
high-pressure washing 69
hollow tile 45
honeycombing 21, 23, 39
Hot Camera 18
housekeeping procedures 4
hurricane 29, 30
hydrochloric Acid 35, 36, 47
impact-echo test 24
impact. 6, 15, 20, 23, 24, 29, 35, 41, 58
Implosion 6
Impulse Response Spectrum 23
Impulse Response test 41
Infra-Red .... See Thermal Imaging
Inspection 1, 3-8, 14, 17, 25, 31, 32, 37, 39, 41, 43, 47, 49, 51, 53, 58-65, 68-70, 75-77, 83-88, 91, 96, 98, 101, 104, 105, 107, 111, 115, 116, 119
inspection of chimneys and stacks 1
Inspection programs 1, 3
Inspection Team Qualifications 32
Inspection techniques 1
insulation 11, 26, 38, 40, 48, 56-59, 61, 73, 75, 76, 83
insulation failure 26
insulation loss 26
integrity testing 6
interior 6, 11, 15-17, 38, 39, 42, 45, 48-50, 69, 76, 79, 82-85, 93, 101
Internal dampers 56
International Committee on Industrial Chimneys .... See CICIND
IR .... See Thermal imaging
Izod Test .... See ASTM E23
jib cranes 56
jumpform 35
laboratory assessment 1, 6, 14
ladders .. 4, 5, 16, 17, 23, 37-40, 47-49, 69, 75, 76, 84, 85, 92, 110, 114
Lagging 43, 57, 59, 62, 88
lead-based paints 5

Lightning protection system 38, 42, 48, 52
Lightning Protection/Grounding System 92, 94
Liner 69-73, 78, 82, 94, 98, 102
liner interior 5, 15, 31, 67
linings 4, 26, 47, 55, 73, 89, 90, 119
Liquid Collectors 92, 94
load cycling 36
long term monitoring program 6
man-lifts 37
masonry chimneys 27, 45
Mass Damper 94
Measuring Thickness by Manual Ultrasonic Pulse-Echo ..... See ASTM E797
microcracking 21
Microwave Equipment 92, 94
modal analysis 29, 30
moisture content 11, 14, 22, 26, 68
mortar 11, 27, 28, 45-51, 67-70
Mortar joints 48, 49, 50
multi flue 71
multi-flued chimneys 7
NACE 90, 118
National Association of .... Corrosion Engineers See NACE
National Institute for Occupational Safety and Health 5
NDT 18, 20, 94
NIOSH 5
Nondestructive 6, 18, 94
nondestructive tests 1, 14
Notched Bar Impact Test .... See ASTM E23
nuclear plants 7
Occupational Safety and Health Administration .... See OSHA
off-line 5, 6, 7, 15, 31, 55, 76
on-line 5, 55, 61
operating conditions .. 3, 4, 7, 8, 35, 47, 67, 89
OSHA 5, 33
OSHA 29 CFR 1926 33
outage schedule 4

Overheating 6, 58
owners 1, 3, 4, 92
Painting 55, 89, 118
Particulates 94
peeling paint 16
penetration...11, 18, 19, 21, 28, 38, 40, 48, 50
Penetration Resistance of Hardened Concrete................ See ASTM C803
permeability 28, 37
Petrographic analysis........................28
petrographic examination 41
Petrographic examination................51
pH 19, 37
Photographs.................. 1, 17, 96, 105
pitting corrosion...............................36
Platform Lighting..........................94
platforms4, 5, 7, 16, 23, 31, 37, 39, 47, 49, 69, 75, 76, 84, 85, 92-94, 104
plumbness of the liner 69
popout of aggregate.........................35
Portland cement............. 35, 46, 47, 67
Ports 62, 88, 94
positive pressure 7, 35
positive pressure gas flow 7
potential difference............ See half-cell potential
precast concrete...............................35
Precipitator 94
Pressurization System 92, 95
Pressurization Systems....................80
prioritizing and planning maintenance needs 7
process plant chimneys......................7
Profile 95
QA/QC 83
qualified person.................................4
Quench Systems...............................80
radial brick............................ 45, 47, 71
rainhood 38, 40, 41, 48, 56, 76, 80, 84, 92, 94
Rainhood 95
Rapid temperature transients 74
rebound number .... See ASTM C805
record keeping 31

Refractories 55
reinforcing bands 16-18, 27, 68, 69, 70, 104
repair... 3, 6, 32, 40, 44, 54, 82, 96, 97, 105, 116
report archiving .... See Record keeping
Request For Proposal......................31
resistivity 22, 23
Roof flashing 56
roughening.......................................14
Safety 5, 32, 42, 52
Saf-T Climb 56
Schmidt Hammer 20
scope of the inspection......................1
Scrubber......................................95
scrubber leaks..................................68
scrubber systems.........................6, 66
Service life prediction 19
severe environmental conditions.........3
shell. 3, 5, 15-17, 35-41, 45, 47-51, 56, 58, 66, 67, 69, 70-72, 75, 76, 78, 80, 83, 84, 91-95, 102, 103, 112
Shell..95
Shell Internal Lining........................60
Shell Opening Reinforcement 95
Shell Plate................................. 59, 63
Shell/Lining Caps..........................95
signatures ...... See structural parameters
Silencers..56
SIR..See GPR
slipform 35, 43
Sonic Mobility test............ See Impulse Response Spectrum
spalled concrete 39, 48, 49
Special coating systems...................55
Splice Bolts 59
spotter scope 5, 16
SSPC.. 89, 118
stability............................... 3, 18, 56
Staining 39
stainless steel 49, 73, 74
stairs..4
standards or codes of practice 18
Stayrods/Bumpers Guides................80
steel chimney 71

steel flue duct systems 25
steel liners 25, 71-75, 78, 81
steel reinforcing bands 27
steel stack 55, 56, 58
Steel Stack Flues 57
Steel Stack Shells 56
Steel Stack Support 55
Steel Stack Types 55
Steel Structures Painting Council ... See SSPC
Strakes 95
structural integrity 3, 15, 29, 30, 34, 97, 105
structural modifications 3, 7, 35
structural parameters 7
structural review 3
Subsurface Interface Radar 26, See also GPR
sulfuric acids 35, 47, 75
sulfuric acid 35
sulfuric acid 47
sulfurous acids 75
swing-stage 93
Swiss Hammer ... See Schmidt Hammer
temporary platform 16
tensile strength 36
The Corrosion Society .... See NACE
The Inspection of Coatings And Linings 89, 118
thermal gradients 34
Thermal imaging 26
thermal stresses 11, 34, 36, 82
Thermally induced movement 68
Thickness of Concrete 119
Thimble 95
titanium alloys 73

Top Cone .... See Exit Cone
Top Suspended Liner 71
toxins in ash deposits 5
trade names 2, 95
Transient Dynamic Response .... See Impulse Response Spectrum
tricalcium aluminate .... See Portland cement
turbulent airflow 26
Turning Vanes 56, 92, 95
Ultrasonic Pulse Velocity .... 21, See ASTM C597, See ASTM C597
Ultrasonic Testing 25
Ultrasonic Thickness Gage 25, 26
ultrasonic thickness gauge ... 75, 77, 83, 85
Ultrasonic thickness measurements . 41, 51, 77, 85
unplanned expenditure 3
unplanned outages 3
Unusual Occurrence 58
UT .... See Ultrasonic Testing
vent stacks 7
vibration 3, 82
visual inspection...5, 15, 18, 38, 39, 47, 49, 59, 68-69, 75-77, 83-85, 102
Visual inspection 1, 14
Visual Inspection 37
vortex shedding 6, 11
Water tables 48
wet precipitation 6
Whittemore Gage 24
wind loads 3, 34, 36
Windsor Probe Test .. See ASTM C803
written report 8